Numerical experiments over the last thirty years have revealed that simple nonlinear systems can have surprising and complicated behaviours. Nonlinear phenomena include waves that behave as particles, deterministic equations having irregular, unpredictable solutions, and the formation of spatial structures from an isotropic medium.

The applied mathematics of nonlinear phenomena has provided metaphors and models for a variety of physical processes: solitons have been described in biological macromolecules as well as in hydrodynamic systems; irregular activity that has been identified with chaos has been observed in continuously stirred chemical flow reactors as well as in convecting fluids; nonlinear reaction diffusion systems have been used to account for the formation of spatial patterns in homogeneous chemical systems as well as biological morphogenesis; and discrete-time and discrete-space nonlinear systems (cellular automata) provide metaphors for processes ranging from the microworld of particle physics to patterned activity in computing neural and self-replicating genetic systems.

Nonlinear Science: Theory and Applications will deal with all areas of nonlinear science – its mathematics, methods and applications in the biological, chemical, engineering and physical sciences.

Nonlinear science: theory and applications

Series editor: Arun V. Holden, *Reader in General Physiology, Centre for Nonlinear Studies, The University, Leeds LS2 9NQ, UK*

Editors: S. I. Amari (Tokyo), P. L. Christiansen (Lyngby), D. G. Crighton (Cambridge), R. H. G. Helleman (Houston), D. Rand (Warwick), J. C. Roux (Bordeaux)

Control and optimization   J. E. Rubio
Chaos   A. V. Holden (*Editor*)

Other volumes are in preparation

Automata networks in computer science

Theory and applications

Edited by Françoise Fogelman Soulié, Yves Robert and Maurice Tchuente

PRINCETON UNIVERSITY PRESS
PRINCETON, NEW JERSEY

Published by Princeton University Press,
41 William Street,
Princeton, New Jersey 08540

Library of Congress cataloging in publication data

Automata networks in computer science:
  (Nonlinear science)
  Includes index.
  1. Cellular automata.   2. Computer networks.
  I. Fogelman Soulié, Françoise, 1948–
  II. Robert, Yves.   III. Tchuente, Maurice,
  1951–   .   IV. Series
  QA267.5.C45A98   1987   006.3   87-42736
ISBN 0-691-08479-3 *hardcover*

Phototypeset in Hong Kong
by Graphicraft Typesetters Limited

Printed in Great Britain
by Biddles Ltd., Guildford and King's Lynn

Contents

Preface

This volume originated from the Workshop 'Computation on Cellular Arrays: theory and applications' which was organised in Paris on 19–20 September 1985, by Françoise Fogelman Soulié, Yves Robert and Maurice Tchuente.

The workshop was sponsored by the Centre National de la Recherche Scientifique (CNRS), in the framework of the Action Thématique Programmée 'Outils et Modèles Mathématiques pour l'Automatique, l'Analyse des Systèmes et le Traitement du Signal' of the Département des Sciences Physiques pour l'Ingénieur.

The contributors of this volume acknowledge financial support from:

ATP 'Outils et Modèles Mathématiques pour l'Automatique, l'Analyse des Systèmes et le Traitement du Signal' du CNRS.
Projet de Recherches Coordonnées 'Mathématiques et Informatique' du CNRS.
Projet de Recherches Coordonnées 'Coopération, Concurrence et Communication' du CNRS.
Fondo Nacional de Ciencias, Chile.

and

Centre d'Etude des Systèmes et Technologies Avancées, Paris.
DIB Universidad de Chile.
Institut National Polytechnique de Grenoble.
Laboratoire de Dynamique des Réseaux, Paris.
Laboratoire TIM3, Grenoble.

Grenoble, July 1986

List of contributors

Michel Cosnard   CNRS, Institut National Polytechnique de Grenoble, and TIM3**

Jacques Demongeot   Université Scientifique, Technologique et Médicale de Grenoble, France, and TIM3**

Françoise Fogelman Soulié   Laboratoire LIA, EHEI, Université de Paris 5, 45 rue des Saints Pères, 75006 Paris, France, and LDR*

Patrick Gallinari   CNAM, 292 rue Saint Martin, 75003 Paris, France, and LDR*

Eric Goles   Dept. Matematicas, Esc. Ingeniería, Universidad de Chile, Casilla 170, Correo 3, Santiago, Chile, and TIM3**

Yann Le Cun   Laboratoire IAAI, Département d'Informatique, ESIEE, 89 rue Falguière, 75015 Paris, France, and LDR*

Driss Moumida   Institute National Polytechnique de Grenoble, France and TIM3**

Patrice Quinton   CNRS, IRISA, Campus de Beaulieu, Avenue du Général Leclerc, 35042 Rennes Cedex, France

François Robert   Institut National Polytechnique de Grenoble, France and TIM3**

Yves Robert   CNRS, Institut National Polytechnique de Grenoble, France and TIM3**

Maurice Tchuente   Université de Yaoundé, Département d'Informatique BP 812, Yaoundé, Cameroon and TIM3**

Sylvie Thiria   CNAM, 292 rue Saint Martin, 75003 Paris, France, and LDR*

*Laboratoire LDR, c/o CESTA, 1 rue Descartes, 75005 Paris, France
**Laboratoire TIM3, BP 68, 38402 Saint Martin d'Hères Cedex, France

Françoise Fogelman Soulié and Yves Robert

Introduction

0.1 Historical review

During the last decade, the search for increased computer power has stimulated the development of 'super-computers'. Their architecture is very different in its design from the classical structure introduced by von Neumann [64] in the 1940s. The so-called 'von Neumann architecture' is based on a sequential mode of computation: at any time, only a few components in the machine are active. For a long time, this bottleneck [37] has limited the performance of sequential computers. However, in the last ten years, the introduction of parallel architectures and the use of dedicated hardware implemented on VLSI chips has allowed a drastic increase of computation speeds.

Yet, mastering the parallel functioning of thousands of interconnected components raises important theoretical problems:

how to *analyse* the performance of a given architecture
how to *synthesise* an architecture for a given task
how to make the most of the new technologies (VLSI)

By the end of the Second World War, J. von Neumann, who had been involved in the design of the first computers, proposed to develop the theory of *automata networks* as a mathematical tool to study computers: 'understanding their mathematical structure will prepare us eventually to go further into the almost unexplored theory of parallel computers' [58].

0.1.1 *The self-reproducing automaton*

Cellular automata were introduced in 1948 by J. von Neumann to compare the 'natural' automata (living beings) and the 'artificial' automata (computers). In an attempt to 'abstract the logical structure of life' [64], he tried to develop an automata network that could exhibit one of the major features of life: auto-reproduction. Artificial machines are usually expected to manufacture objects less complex than themselves, unlike natural organisms, which, by means of evolution and adaptation, have continually increased the complexity of their offspring.

The model developed by von Neumann – the self-reproducing automaton [64] – was based on a cellular array: a 2-dimensional grid. Each cell of the grid can be in one of 29 different states, which have been defined in analogy with the nervous system: quiescent or excited state, reception or transmission of the nervous signal, and so on.

Initially, the grid contains two sets of non-quiescent cells: the 'constructor' (200 000 cells) and the 'organism' which is to be reproduced. A dynamical evolution is then run on the network which stops when a copy of the organism is displayed in another part of the grid: the organism has been reproduced.

0.1.2   The game of Life

The universal computation property can also be obtained by a cellular array called Life, which was introduced in 1970 by J. Conway [12]. Life has been designed to simulate a population of interacting living organisms or cells. The population is supposed to 'live' on a 2-dimensional grid (potentially infinite). At each node of the grid is a cell, which can be in one of two possible states: dead or alive (0 or 1). Only a finite number of cells are alive at any given time (this is called the 'finite-configuration assumption'). The cells change state, in parallel, by applying, each for its own sake, a transition function which depends only on the 8 nearest neighbours of the cell on the grid (Moore neighbourhood: Fig 0.1).

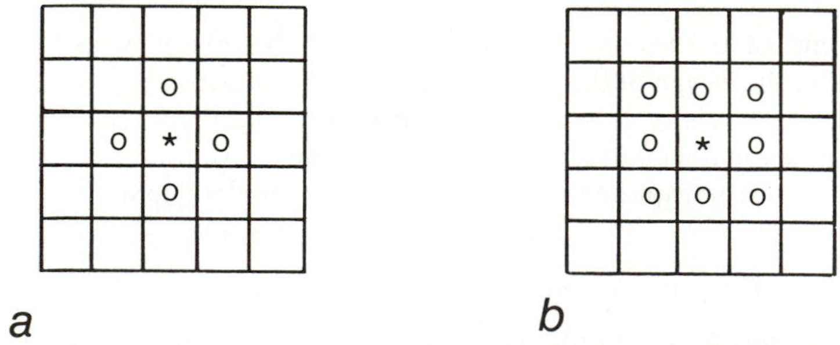

a   *b*

Figure 0.1 Neighbourhoods on a 2-dimensional grid. The figure shows the neighbours (○) of a cell (∗) in the von Neumann neighbourhood (a: 4 neighbours) and the Moore neighbourhood (b: 8 neighbours).

The transition functions are chosen so as to 'mimic life':

death: a living cell can survive for the next generation if and only if it has exactly 2 or 3 living cells in its neighbourhood. Otherwise, it dies.
birth: a cell will come to life at the next generation if and only if it has exactly 3 living cells in its neighbourhood.

Life has been widely described in scientific journals [26], because it exhibits an astonishing variety of dynamical behaviours. Starting from an initial configuration, that is, a given finite set of living cells, we update the state of the population in the following way: at each pulse of a clock, all cells look around and update their state according to the above laws. The *global* behaviour that arises from such dynamics may be (Fig. 0.2):

fixed points
limit cycles
'travelling' configurations
configurations that increase in size: the most famous is the 'glider gun', which produces a glider (Fig. 0.2) every 30 generations.

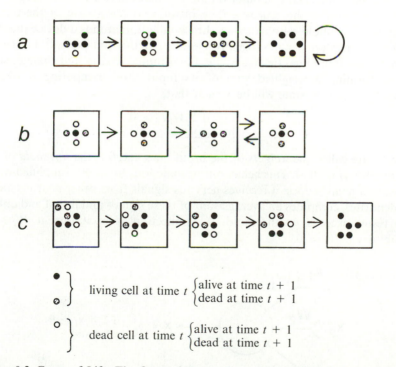

Figure 0.2 Game of Life. The figure shows various dynamical evolutions of finite configurations: (a) a fixed point; (b) a limit cycle of period 2 (flip-flop or traffic light); (c) a travelling configuration (the 'glider'): the initial configuration (with 5 living cells) is found translated 1 cell diagonally after 4 generations.

The interesting feature of Life is that it succeeds, despite its very simple *local* transition rules, to achieve a *global* behaviour, that is a nicely balanced compromise between the complexity of the dynamical evolutions and their stability. There is neither explosion nor implosion of the living population, but an interesting (and fascinating to many scientists!) evolution of the population. Similar versions of Life have been proposed which exhibit this same property [83].

Conway has proved that Life is sufficiently powerful to be a universal computer: he exhibited [12] configurations that realise the usual logical gates (NOT, AND, OR), thereby proving that Life can simulate a Turing machine. See Chapter 6 for further details. Many cellular automata, such as Life, have been studied [20] and proved to be capable of very complex behaviour as well as being a very versatile modelling tool (in physics, biology, and so on).

0.1.3 *The perceptron*

In 1943 [55], W. S. McCulloch and W. Pitts proposed a different model to implement a universal computer on an automata network. In analogy with the nervous cell – the *neuron* – they introduced the notion of the formal neuron or threshold automaton. A threshold automaton is a device that has two possible internal states (usually 0 and 1: inactive and excited). It receives signal along input lines (let $x_1, \ldots, x_n$ be its input signals), and changes state by performing a weighted sum of its inputs and comparing it with a threshold. Its next state will be x such that:

$$x = \begin{cases} 1 & \text{if } \Sigma_j W_j x_j \geq \theta \\ 0 & \text{otherwise} \end{cases}$$

The W_j are called the *weights* of the input lines and θ is the *threshold* of the automaton (Fig. 0.3). This behaviour is analogous (in a very simplified way!) to what a neuron does: it receives nervous signals from other neurons along its dendrites, computes an average sum of these signals and fires if and only if the result exceeds its threshold. A connection is excitatory if $W_j > 0$, inhibitory if $W_j < 0$.

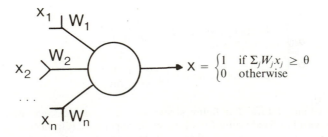

Figure 0.3 Threshold automaton

McCulloch and Pitts have shown [55] that a network made of these very simple automata could simulate a Turing machine.

Rosenblatt [76] and Minsky and Papert [58] made further use of these automata to propose a learning machine: the perceptron. The perceptron is a threshold automata network, with connection weights W_{ij} among automata, that can receive input from the outside on a 'retina'. A learning algorithm was proposed [58] which allowed the perceptron to automatically categorise classes of patterns. The first positive results gave rise to excessive expectations, which were later on almost completely extinguished by the result of 'limitation' that Minsky and Papert established, giving a full mathematical framework to deal with learning machines.

It is only recently that this line of research has started again, under the influence of various factors: the development of mathematical tools, the recent interest of physicists in the theory of disordered matter, new insight in brain functioning and, essentially, the evolution in computer science which has made available fast computers on which to simulate automata networks such as the perceptron.

In the following sections we review these last points: the mathematical theory of automata networks (Section 0.2.1), the use of automata networks in physics (0.2.2), in biology (0.2.3), and, in greater detail, in computer science: theory of computability (0.3.1), design of systolic arrays (0.3.2) and artificial intelligence (0.3.3). Section 0.4 contains a detailed presentation of the subsequent chapters.

0.2 Computing on automata networks

0.2.1 *Definition*

An *automata network* may be defined, in a general way, as a (large) set of cells (finite automata), locally interconnected, which can evolve at discrete time steps through mutual interactions.

From a mathematical point of view, automata networks are *discrete dynamical systems:* space, time, and the cells themselves are basically discrete. They can provide an interesting alternative solution to the very classical and traditional models based on *continuous dynamical systems*, that is, ordinary differential and partial differential equations.

Formally, an automata network can be described as a mapping F from S^n into itself, where S is a finite space (state space). The network is then made up of n interconnected cells. The connection structure is defined by F: cell i receives a connection from j if F_i depends on the jth variable (where F_i is the ith component of mapping F).

A state of the network is a vector x in S^n. A dynamics on the network is then defined through a rule that transforms any vector x in S^n into a vector y in S^n. For example, the parallel iteration rule is defined as:

$$y = F(x)$$

and can be interpreted as follows: at each time step, each automaton computes its next state by using its mapping F_i on the current state x. Different iteration modes may of course be defined.

As soon as a dynamics – or time evolution – of the network is defined, the problem of its asymptotic behaviour arises. As the state space S is usually finite, all trajectories will be periodic: limit cycles, fixed points.

The theory of discrete iterations [70, 71] is one of the tools available to characterise the dynamical behaviour of automata networks. In some particular cases [29], tools from statistical physics [35] have proved useful to give interesting results on the limit cycles and transient lengths of threshold networks. In the more general case of boolean networks [38], few exact results on periods and limit cycles are known. However, the geometrical structure of the limit cycles has been studied in detail [21], and tools, introduced in discrete iteration theory [71] or graph theory, have allowed an approximation or 'average' of the limit cycles.

The model discussed so far is fully deterministic: the transition rule – for example, F in the parallel iteration mode – allows the prediction with full certainty of the next state from the present state. Random automata have also been introduced more recently [65, 68] and their relation to random Markov fields [16] made explicit. They can be used to model renewal processes, epidemiology or problems in computer science such as synchronisation or the restoration of images.

0.2.2 *Physics*

It is only very recently that physicists have been concerned with disordered systems. Until then, they dealt mainly with systems made of identical elements (e.g. gas, crystals) and successfully developed tools that allowed them to establish the classical laws of mechanics and thermodynamics. In the classical approach discrete systems, such as ensembles of particles, have been modelled by continuous variables, leading to ordinary or partial differential equations. Directly producing a discrete model in this case might be simpler and more natural. Sure enough, solving these equations requires numerical discretisations (but the equations were obtained by forcing discrete quantities Δx, Δt, ... to 0!) and floating-point representations of the reals. This introduces discretisation errors, rounding errors and time-consuming floating-point operations.

Since the mid-seventies fully discrete models have been extensively studied: in percolation models, spin glasses and fluid mechanics. Simulation of discrete models on a computer can be performed exactly and quickly, as it operates on bits, which allows the finding of 'experimental' solutions when no exact derivation is possible. Cellular automata, for example, [87, 91] can be implemented on a dedicated machine [85].

The *spin glass* model [41] is a classical model of the thermodynamic properties of disordered systems. Particles are represented as Ising spins ($+1$ or -1) and located at the nodes of a grid (that may be 1-dimensional, 2-dimensional, ..., n-dimensional). The interactions among spins are random. These systems have many different low-energy configurations – or ground states – which are their stable states. The characterisation of these ground states is a complex combinatorial problem.

A general algorithm, known as *simulated annealing* [42] has been proposed to solve such problems of combinatorial optimisation. Derived in analogy with the Monte Carlo method used to find the ground states of a spin glass, it has been successfully applied to a variety of problems [7].

The spin glass model has also been proposed to model *neural networks* [1, 35, 66]. Hopfield [35] proposed to use such networks to memorise a set of predefined patterns and retrieve them from partial or noisy information. The concept of *energy*, analogous to the spin glass energy, then allows the formalisation of this task as the problem of 'shaping' an energy landscape, the memorised items being the low-energy wells of this landscape. The tools of statistical physics have been used to characterise the memorisation abilities of these networks.

Interestingly enough, the spin glass model has proved formally equivalent to the already discussed threshold network, which has allowed the derivation, within the threshold automata framework, of results concerning the dynamical behaviour [21, 29] of neural nets and their memorisation capacity [24, 89].

Similar ideas have also been applied in the context of boolean networks [38]: by comparing a boolean network with an annealed model (where at each time step, the mapping F is set at random), it has been possible to predict the time evolution of the overlaps between configurations [18]. This result is to be compared with the results obtained in automata theory about the stable core of limit cycles [22, 23].

The recent development of discrete models in fluid mechanics to model turbulence has proved competitive with the traditional finite elements approach of Navier–Stokes partial differential equations [36]. The simulation of the model on a computer is fast and implementation on a dedicated chip will further reduce this computation time. Work on the design of such chips is currently under progress [25].

0.2.3 *Biology*

Automata network based models have been proposed in various fields of biology [14, 15, 30]. We will mention only briefly some of them.

Random boolean networks have been used to model the genetic [38] and immune systems [84]. As the exact interaction structure of the genes is not yet fully understood, Kauffman proposed to model the genetic system by a *random* boolean network where binary deterministic automata are connected in a random way. The dynamical evolution of the network produces limit cycles whose number and periods can be compared with characteristic features of the genetic system of various species [38]: in fact the *scaling laws* relating the number of limit cycles and the average period to the number n of automata in the network vary as $n^{1/2}$, which is also the law relating the number of different cells and replication time to the 'size' of the genome.

This search for generic laws – i.e. independent of the exact wiring of the network – has been pushed forward to provide models of evolution [39]. In particular, it has been shown in [88] that the dynamical behaviour of a boolean network could provide parameters similar to fitness. When one inserts these fitness coefficients into standard differential equation models of population dynamics, the resulting behaviour is that – observed in palaeontology – of punctuated equilibria, a behaviour that these models could not otherwise produce.

The seminal paper of McCulloch and Pitts [55] has been followed by a large number of contributions to brain theory [56], ranging from formal study of the dynamical behaviour of neural networks [13, 24, 29] to implementation of associative memory [19, 35, 60, 89]. Some authors (Edelman, Hopfield) believe in the possibility of building machines, whose design would be based on what is known of the brain and which would share some of the properties of the brain. Many chips – neural networks on silicon – are currently being designed in the USA.

0.3   Automata networks in computer science

0.3.1   *Computability on cellular structures*

Following the pioneering work of von Neumann, the theory of computability on cellular structures was first oriented towards the study of structures simpler than the 29-state 5-neighbourhood cellular automaton presented above, and which are both computation- and construction-*universal*. As explained in Burks [8], such structures can simulate any Turing machine and, given any quiescent automaton, they can construct that automaton in a designated empty region of the cellular space. For instance, Codd [9] has shown that there exists a 2-dimensional 8-state 5-neighbourhood automaton which is computation-construction universal. In addition, Smith [80] has proved that there exist very simple 1-dimensional computation-universal structures which can support nontrivial self-reproduction.

Subsequently, some efforts have been made to solve problems by exploiting the inherent high parallelism of cellular automata. For instance, it is known from Barzdin [6] that the time required to recognise a palindrome of size n in a 1-headed Turing machine is of the order of n^2, but Smith [80] has exhibited a 1-dimensional array with nearest-neighbour interaction that can solve the problem in time $n/2$. Another interesting result is due to Kosaraju [44] who has shown the following complexity result: any context-free language is recognisable in time $(1 + \varepsilon)n$ by a 2-dimensional cellular automaton and in time n^2 by a 1-dimensional cellular automaton.

However, all these studies are closely related to the sequential model of Turing machines. This is well illustrated by the following quotation from Smith [80]: 'for a first approximation of the complexity of cellular automata, one can say that they lie somewhere between the completely serial Turing machine and the highly parallel multi-headed machines, for certain types of problems'.

The first mathematical studies devoted to the characterisation of all the transformations which are realisable on a uniform cellular structure are due to Hedlung [31] and Richardson [69]. They show independently the following result: for any finite non empty set of states Q, and for any integer $n \geq 1$, if the product topology induced by the discrete topology of Q is assigned to the set of configurations $\mathscr{C} = \{C: \mathbb{Z}^n \mapsto Q\}$, then a mapping F from $\mathscr{C}$ onto itself can be realised by a uniform cellular structure if and only if it is continuous and commutes with the shift operation.

Several authors have then gone on to study the problem of determining the capacity for computation of a flexible cellular structure $A = (\mathbb{Z}^n, Q, V)$ where the neighbourhood index is fixed but the transition function may change from one time step to another.

In this context, a mapping $F: \mathscr{C} \mapsto \mathscr{C}$ is said to be computable on A if and only if it can be decomposed into the form $F = F_1 \circ F_2 \circ \ldots \circ F_n$, where each F_i is a possible global transition function of the system.

The first result in this direction is due to Amoroso and Epstein [2], who showed that for $Q = \{0, 1\}$ and for any $n \geq 2$, there exist indecomposable maps, that is global transition functions of a binary 1-dimensional scope-n cellular structure which are not computable on any binary 1-dimensional scope-m cellular structure, with $m < n$. This model of computation is very close to the classical completeness problem where one looks for a complete set of functions [75].

In the case of a finite network $\mathscr{N} = (G, Q)$, where G is the connection graph, Gastinel [27], Tchuente [81], Kim and Rousch [40], have been interested in the computation of linear mappings. In the particular case where Q is a field, Tchuente [82] gives a characterisation of all boolean transformations computable on a tree-connected network of binary automata. When computation on a finite network $\mathscr{N}$ is defined in terms of factorisation of functions, the permutations $\sigma: (x_1, \ldots, x_n) \mapsto (x_{\sigma(1)}, \ldots, x_{\sigma(n)})$ play a central role. Indeed, such functions can be considered as a formal model of the situation where, for any $1 \leq i \leq n$, automaton x_i wants to send a message to $x_{\sigma(i)}$. As a consequence, if $l_G(\sigma)$ denotes the minimum length of factorisation of σ, then $l_G = \max_\sigma l_G(\sigma)$ is the maximum delay necessary for the transmission of collections of messages of the form

$(\sigma(1), \ldots, \sigma(n))$. Such a model seems to be appropriate for measuring the communication capacity of a cellular structure (see Chapter 6).   o

0.3.2   *From iterative arrays to systolic architectures*

Another aspect of computability on cellular structures has been developed by Hennie [32] in connection with the analysis and design of logical circuits. The underlying idea is that iterative networks composed of identical cells interconnected to form a regular array are very suitable for practical realisation. Indeed, such structures are economical to manufacture and repair, and, also, they can be enlarged to accommodate more variables simply by adding more components, that is without changing the existing portion of the network. The fundamental work of Hennie was concerned with the following general questions: 'what kind of operations can iterative networks perform, how can they be analysed, and how can an iterative array be designed to do a specific job?'

However, this work, like that of many other authors in the area, did not come to practical realisation. We see two main reasons for this: first, efforts were engaged in establishing very general results rather than solving specialised problems, and second, the work was largely ahead of techno-logical possibilities.

Today, with the development of VLSI [57], it has become clear that the fundamental aspects of cellular structures, namely *regularity*, *modularity* and *parallelism*, can lead to the concrete realisation of high-performance devices. The major contribution of Kung and Leiserson [49] is to have realised that cellular networks should be conceived not as universal machines (see von Neumann [64]), but rather as procedures on silicon.

Indeed, such parallel computing resources can be efficiently implemented with today's high circuit density VLSI technology, provided that the two following conditions are satisfied:

> the resource is a special-purpose computing device which is attached to a host architecture (rather than a universal machine)
>
> the application run on the resource is a compute-bound application, where the number of arithmetic operations is much larger than the number of input and output elements. Front-end processing in radar, sonar, vision, or robotics yields typical examples of compute-bound applications [48].

Systolic arrays, as defined by Kung and Leiserson [49], are a useful tool for designing special-purpose VLSI chips. A chip based on a systolic design consists of essentially a few types of very simple cells which are mesh-interconnected in a regular and modular way, and achieves high performance through extensive concurrent and pipeline use of the cells.

The name *systolic* given by Kung and Leiserson is taken from the physiology of living beings. A systole is a contraction of the heart, by means of which blood is pumped to the different components of the organism. In a systolic system, the information (data and instructions) is rhythmically sent through a structure of elementary processors (cells), to be processed and passed to neighbouring cells until a result reaches some border of the system communicating with the host [62].

To establish a connection between systolic architectures and cellular automata, we can say that systolic algorithms are algorithms which may be efficiently realised on systems that are representable as finite deterministic time-invariant synchronous cellular automata [62, 63].

Kung carefully explains [46, 47] how systolic architectures should result in cost-effective and high-performance special-purpose systems, and he gives three main reasons to justify the advantages of the systolic model:

systolic systems are easy to implement because of their simple and regular design

they are easy to reconfigure because of their modularity

they permit multiple computations for each memory access, which speeds up the execution without increasing I/O requirements.

Such regular networks of tightly coupled simple processors with limited storage have provided cost-effective high-throughput implementations of important algorithms in a variety of areas (signal and image processing, speech and pattern recognition, matrix computations), see [37, 45, 47, 48, 50, 51, 54, 72, 73, 77, 85] among others. It seems important to provide designers with methods that help them explore various implementations of the same algorithm. Several alternatives may then be compared according to various criteria such as the size of the array, the complexity of the elementary cells, the throughput and the pipelining delay. Attempts to synthesise systolic arrays can be classified into three categories.

The first approach, called *functional transformations*, consists of applying formal transformations to the mathematical expressions of the algorithm [10, 88]. These transformations introduce timing considerations through the use of the delay operator z^{-1}.

A second approach, called *retiming*, is due to Leiserson and Saxe [53]. This method consists of applying a graph transformation to a design that is not systolic (in the sense that the data may have to go through an arbitrary number of combinatorial logic operations at each cycle) to obtain an equivalent systolic design. See Section 8.5 for a description of this method. However, note that the retiming method fails to capture the geometric regularity of systolic arrays, since it can be applied to an arbitrary design.

A third approach called *dependence mapping*, aims at extracting the dependences between the variables of a program and mapping the program onto a systolic array in such a way that the dependences are preserved. Attempts in this direction have been recently proposed by Moldovan [61], Miranker and Winkler [59], and Quinton [67] (see Chapter 9).

0.3.3 *Automata networks for artificial intelligence*

In the 1950s J. von Neumann initiated the work on automata networks with the goal of understanding the logical differences between natural and artificial 'intelligence'. The problem was further developed by Minsky and Papert [58] in their study of learning machines. The general problem at hand was that of *learning from examples:* examples are presented to the machine, which is supposed to adapt its structure so as to be able to retrieve the examples later on. Usually fault-tolerance is required in addition, which means that noisy patterns must allow for correct retrieval of stored items.

The 'perceptron limitation theorem' [58] stated – roughly speaking – that perceptrons could only learn to solve 'easy' problems. It was widely understood as establishing the failure of this line of research.

But Minsky and Papert were quite aware of the restrictions they had imposed on their model: the perceptron has only one layer of adaptive parameters (the connection weights) and makes use of threshold automata. Now, the limitations of the threshold automaton are well known: it can only classify ensembles which are linearly separable. Using more general functions would probably help in reaching a solution in 'hard' problems. On the other hand, introducing more than just one layer of adaptive weights seems very appealing: usually one expects increased performances from a larger set of available parameters. The problem is then obviously to design a learning rule. These two solutions: random automata and multi-layered networks have recently been implemented in the Boltzmann Machine [34] and it has been proved that learning machines could be developed to solve 'hard' problems (see Chapter 7 for more details).

This new approach – called *connectionism* [34] – is based on the use of automata networks. A learning rule allows the network to progressively update its connection weights so as to encode the examples. The resulting representation of knowledge is *distributed* and compact. The retrieval process consists in iterating the network, which can allow for fast *parallel* updating. Parallel machines, such as for example the Connection Machine [33] are now being used to test some applications.

The main outcome of this latest effort in the area is the design of a general learning rule, the Gradient Back Propagation algorithm, for multi-layered networks [52, 78]: it provides a general method by which an automata network can extract information from examples, store it using its own internal representations and finally use it to perform 'hard' problems. Various applications are under way in such areas as image processing, speech and pattern recognition, expert systems and natural languages.

0.4 Outline of the book

This book is intended to provide a general introduction to the field of automata networks, focusing mainly on the theoretical tools (section 0.4.1) and some applications in computer science (section 0.4.2). Of course, we do not claim to cover the whole field. The interested reader may refer to Section 0.5 for further references.

0.4.1 *Mathematical tools*

Chapter 1 is an introduction which presents some basic notions in discrete iteration theory. The main definitions of local transition function, operation modes, iteration graph and limit cycles are given and illustrated by some simple examples. Then, we define the concepts of boolean distance and discrete derivative, allowing the characterisation of the property of local attractivity of fixed points. All the tools presented here may be considered as the implementation in the discrete case of the corresponding notions in the continuous case.

In the appendix of the chapter, we give a characterisation of the geometrical structures of the limit cycles in boolean networks. Those limit cycles have in common a set of elements which remain always fixed – the stable core. An algorithm is given to approximate this stable core.

In Chapter 2, we establish a connection between continuous and discrete systems through the theory of itineraries. A basic aspect in the study of the iterative behaviour of 1-parameter families of unimodal functions is the combinatorial character of various properties, which is investigated using coding techniques.

We introduce a simplified theory of itineraries in order to study the relationships between the iterative behaviour of a selfmap and an associated dynamical discrete system. In the following sections, we concentrate on unimodal functions and continuous increasing selfmaps of the circle.

In Section 2.5, we go from itineraries to automata, and study the dynamical behaviour of the following automaton with memory:

$$y_{n+1} = \mathbf{1}\left(\sum_{i=0}^{p} \alpha_i y_{n-i} - \beta \right)$$

where y_n belongs to $\{0, 1\}$, α_i and β are real parameters, p is the size of the memory, and the function $\mathbf{1}$ is defined as $\mathbf{1}(u) = 1$ if $u \geq 0$, 0 otherwise.

The case of a bounded memory (p constant, independent of n) is studied in Chapter 5. Here, we restrict ourselves to the case of an unbounded memory: $p = n$, and $\alpha_i = \alpha^i$ where $0 < \alpha < 1$ (β is arbitrary). We show that there always exists a unique attractor, which is either a cycle or a Cantor set.

In Chapter 3 we introduce the general definition of random automata.

Three examples of such networks are presented and their dynamical properties studied through the use of an energy or entropy function. Simulations are shown to illustrate the case of a renewal process. The random automata network model presented here also allows us to implement a Gibbs sampler.

In Chapter 4, the general tool of Lyapunov – or energy – function is introduced and applied to threshold networks. The dynamical behaviours of various modes of iteration on threshold networks are illustrated through simple examples. Then, for each of the parallel, sequential and block-sequential modes of iteration, the corresponding energy function is defined. We then show that the use of this function allows us to characterise the dynamical behaviour of the network. In particular, conditions are given on the connection matrix of the network which ensure that the network will have only fixed points (sequential and block-sequential iterations) or cycles of period at most 2 (parallel iteration). Furthermore, we derive bounds on the transient length.

Chapter 5 is a case study; we study the dynamical behaviour of the same automaton as at the end of Chapter 2, but with the restriction of a bounded memory. The equation of the boolean automaton is now:

$$x_{n+1} = \mathbf{1}\left[t + \sum_{i=0}^{k-1} a_i x_{n-i} \right]$$

where x_n is the state of the automaton at the discrete time step n, a_i are real coupling coefficients, t is the threshold; k is the size of the memory, it is an integer constant.

The general dynamic of such a model is extremely rich. We present some results concerning the particular cases in which the coefficients a_i form a geometric sequence and a palindromic word. Then we characterise the reversible automata according to the values of a_i and t.

We conclude this part of the book by a general overview of the theory of computability on automata networks (Chapter 6). First we give three examples of infinite uniform deterministic cellular structures, including the game of Life. In connection with Life, we address in Section 6.2 the important problems of universal construction and self-reproduction.

Synchronisation problems (Firing Squad and self-stabilisation) are briefly mentioned in Section 6.2.3. Then we deal with two approaches for the notion of computability on uniform infinite cellular structures: first we show how to simulate a Turing machine with a 1-dimensional cellular automaton [80]; in the second model of computation, the global transition function can change during time, and the problem is to characterise the mappings which can be decomposed into a product of admissible transition functions.

In Section 2.5, we concentrate on finite networks $\mathcal{N} = (G, Q, \mathcal{F})$ where G is a directed graph of order n representing the interconnection structure, Q is the finite nonempty state set, and $\mathcal{F}$ is a collection of functions from Q^n

into itself, representing the set of possible global transition functions of the network.

First we characterise the computable functions in flexible networks where the constraints on the local transition functions are only due to the interconnection structure. Then we deal with tree-networks, where the interconnection graph G of the network is a tree and $\mathcal{F}$ is the set of all mappings from Q^n into itself. Finally, we study the minimum factorisation length of the transition functions, and we give a criterion for the comparison of trees with respect to the factorisation of boolean permutation matrices.

0.4.2 *Applications*

Chapter 7 is a general review of the use of automata networks in artificial intelligence, and more precisely in learning and the distributed representation of knowledge. We first present an overview of the classical adaptive machines: linear classifiers, adaline, perceptrons, and of the related learning algorithms.

We then develop the theory of linear learning [43] and discuss the different computational techniques. Learning examples, in this framework, amounts to computation of the 'best' connection matrix of the network. This connection matrix is a solution of a matrix equation that can be solved by using the theory of generalised inverses. The solution is usually not unique.

Various modes of retrieval are then introduced. For the linear retrieval process, we discuss the optimality of the connection matrix computed by the learning process for various models of noise. We then discuss the threshold retrieval process, applying tools developed in Chapter 4. An example of pattern recognition is used to compare performances with other retrieval modes: 'Brain State in the Box' model – BSB – and smooth.

We show for this learning rule and the different retrieval processes that it is possible to automatically produce networks capable of learning examples and retrieve them from noisy input. However, these techniques appear to suffer from intrinsic limitations, just in the same way as the perceptrons did.

In the last section of the chapter, we develop the multi-layer network model to go beyond this limitation. The problem there is to design a learning rule by which the connection weights of internal cells (in intermediate layers) can be updated. We introduce the Gradient Back Propagation algorithm and show that it can provide an efficient learning rule by back-propagating error signals through the network. Its possibilities are illustrated by a few examples.

Chapter 8 is deveoted to the design and analysis of systolic arrays. First, the basic concepts of the systolic model that we have introduced in Section 0.3.2, are illustrated with some simple examples. The first is the design of a systolic array for the 1-dimensional convolution problem (nonrecursive and recursive). The most general instance of the convolution problem, or IIR filtering, is the following [47]: given a sequence of input signals $(x_i)_{i \geq 1}$, compute

$$y_i = a_1 * x_i + a_2 * x_{i-1} + a_3 * x_{i-2} + \ldots + a_k * x_{i-k+1}$$
$$+ r_1 * y_{i-1} + r_2 * y_{i-2} + r_3 * y_{i-3} + \ldots + r_h * y_{i-h}$$

The array presented in the text [47] solves this problem using $h + k$ inner product step cells and a delay cell. A new y_i is delivered every second time-step, and the cycle delay depends neither on the length k of the filter, nor on the order h of the recurrence.

A systolic array for matrix-vector multiplication is introduced in Section 8.2.3, and the solution of triangular linear systems is dealt with in Sections 8.2.4 and 8.2.5 [49], [54]. Again, linear computation times are obtained when using a 1-dimensional linear array.

In Section 8.3, harder examples are considered. Let **A** be a dense $n \times n$ matrix, and **b** be a vector with n components. To solve the linear system $\mathbf{Ax} = \mathbf{b}$, we first have to triangularise the matrix $(\mathbf{A}, \mathbf{b})$ on a 2-dimensional systolic array, using either Gaussian elimination or Givens factorisation [28]. Once the factorisation $\mathbf{A} = \mathbf{LU}$ (Gauss) or $\mathbf{A} = \mathbf{QR}$ (Givens) is obtained, there is a triangular system to be solved; this can be achieved using the array of Section 8.2.5. A 1-array approach for the direct solution of the linear system of equations $\mathbf{AX} = \mathbf{B}$, where **X** and **B** are $n \times p$ matrices, is depicted in Section 8.3.2: [73], [74]. This array implements the Gauss–Jordan diagonalisation algorithm. These examples should give the reader a good insight of what type of problems could be efficiently solved using systolic architectures.

The advantages of the systolic model are investigated in further detail in Section 8.4. The last section of the chapter is more theoretical. Indeed, in Section 8.5, we present the methodology of Leiserson and Saxe [37] which aims to systematically derive systolic architectures from synchronous circuits.

As already mentioned, the retiming methodology of Leiserson and Saxe is not properly speaking a synthesis method, as the starting point is a particular implementation of the algorithm whose basic properties are not altered by the retiming. On the contrary, a systematic method for the design of systolic arrays is described in Chapter 9. The method is informally explained in Section 9.2, through the simple example of the convolution product.

More technically, we use this method for those algorithms that can be expressed as a set of uniform recurrence equations over a convex set D in $\mathbb{Z}^n$. Most of the algorithms already considered for systolic implementation can be represented in this way.

The method consists of two steps: finding a schedule of computations (or timing function) that is compatible with the dependences introduced by the equations, then mapping the domain D (allocation function) onto another finite set of integer coordinates, each representing a processor of the systolic array, in such a way that concurrent computations are mapped onto different processors. The scheduling and mapping functions meet conditions that allow the full automation of the method. In particular, criteria for the design of optimal timing functions are derived (see Section 9.3).

We show (Section 9.4) how the method can be applied to derive new systolic designs, in the case of the convolution product and the matrix product. A generalisation of the notion of uniform recurrence equations is introduced in Section 9.5 in order to handle problems such as two-level pipelining. For the sake of clarity, proofs of the main theorems are given in the appendices.

0.5   References

[1] D. Amit, H. Gutfreund and H. Sompolinsky, Storing infinite number of patterns in a spin glass model of neural networks, *Phys. Rev. Lett.*, 55 (1985), 1530–3

[2] S. Amoroso and I. J. Epstein, Indecomposable parallel maps of tessellation structures, *J. Comput. Syst. Sci.*, 6 (1976), 136–42

[3] M. A. Arbib, *Theories of Abstract Automata*, Prentice-Hall, Englewood Cliffs, NJ (1969)

[4] A. J. Atrubin, A one-dimensional real time iterative multiplier, *IEEE Trans. Comp.* 14, 6 (1965), 394–9

[5] J. Backus, Can programming be liberated from the von Neumann style?, *Comm. of the ACM*, 21, 8 (1978), 613–41

[6] V. M. Barzdin, Complexity of recognition of symmetry in Turing machines, *Problemi Kibernetiki*, 15 (1965)

[7] E. Bienenstock, F. Fogelman Soulié and G. Weisbuch (eds), *Disordered Systems and Biological Organization*, NATO ASI Series in Systems and Computer Science, F20, Springer Verlag, Berlin, Heidelberg, New York (1986)

[8] A. W. Burks, *Essays on Cellular Automata.*, University of Illinois Press, Urbana (1970)

[9] E. F. Codd, *Cellular Automata*, Academic Press, New York (1968)

[10] D. Cohen, Mathematical approach to iterative computation networks, *Proc. 4th Symposium on Computer Arithmetic, IEEE* (1978), 226–38

[11] S. N. Cole, Real-time computation by *n*-dimensional iterative arrays of finite-state machines, *IEEE Trans. Comp.* 18, 4 (1969), 349–65

[12] J. H. Conway, E. R. Berlekamp and R. K. Guy, *Winning Ways for your Mathematical Plays*, Academic Press, New York (1982)

[13] M. Cosnard and E. Goles, Dynamique d'un automate à mémoire modélisant le fonctionnement d'un neurone, *C. R. Acad. Sci.*, 299, 1, 10 (1984), 459–61.

[14] M. Cosnard, J. Demongeot and A. Lebreton (eds), *Rhythms in Biology and Other Fields of Applications.* Lecture Notes in Mathematics 49, Springer Verlag, Berlin, Heidelberg, New York (1983)

[15] J. Della Dora, J. Demongeot and B. Lacolle (eds), *Numerical Methods in the Study of Critical Phenomena*, Springer Verlag, Berlin, Heidelberg, New York (1981)

[16] J. Demongeot, Random automata and random fields, in [17], 99–110

[17] J. Demongeot, E. Goles and M. Tchuente (eds), *Dynamical Systems and Cellular Automata*, Academic Press, New York (1985)

[18] B. Derrida and G. Weisbuch, Evolution of overlaps between configurations in random boolean networks, *Journal de Physique*, to appear

[19] G. M. Edelman and G. N. Reeke, Selective networks capable of representative transformations, limited generalizations and associative memory, *Proc. Nat. Acad. Sci. USA*, 79 (1982), 2091–5

[20] D. Farmer, T. Toffoli and S. Wolfram (eds), *Cellular Automata*, North-Holland, Amsterdam (1985)

[21] F. Fogelman Soulié, Contributions à une théorie du calcul sur réseaux, Thesis, Grenoble (1985)

[22] F. Fogelman Soulié, Frustration and stability in random boolean networks, *Discrete Appl. Maths.* 9 (1984), 139–56

[23] F. Fogelman Soulié, Parallel and sequential computation on boolean networks, *Theoretical Computer Science* (1985)

[24] F. Fogelman Soulié and G. Weisbuch, Random iterations of threshold networks and associative memory, *SIAM J. Comput.* (1986), 1505

[25] U. Frisch, B. Hasslacher and Y. Pomeau, A lattice gas automaton for the Navier–Stokes equation, *Phys. Rev. Lett.* (1986), 1505

[26] M. Gardner, Mathematical games, *Scientific American* (1970), 120–3

[27] N. Gastinel, Réalisation du calcul d'une transformation linéaire aux noeuds d'un graphe, *Colloque sur les méthodes de calcul pour des systèmes de type coopératif (1978)*, Giens, France

[28] W. M. Gentleman and H.T. Kung, Matrix triangularisation by systolic arrays, *SPIE Symp. 298, Real-time Signal Processing IV* (1981), 19–26

[29] E. Goles Chacc, Comportement dynamique de réseaux d'automates, Thesis, Grenoble (1985)

[30] J. M. Greenberg, B. D. Hassard and S. P. Hastings, Pattern formation and periodic structures in systems modelled by reaction diffusion equations, *Bull. Am. Math. Soc.* 84, 6 (1978), 1296–1327

[31] G. A. Hedlund, Endomorphisms and automorphisms of the shift dynamical system, *Math. Syst. Theory* 3 (1969), 320–75

[32] F. C. Hennie, *Iterative Arrays of Logical Circuits*, MIT Press, Cambridge, Mass. (1961)

[33] W. D. Hillis, *The Connection Machine*, MIT Press, Cambridge, Mass. (1986)

[34] G. E. Hinton and J. A. Anderson (eds), *Parallel Models of Associative Memories*, Erlbaum, Hillsdale, (1981)

[35] J. J. Hopfield, Neural networks and physical systems with emergent collective computational abilities, *Proc. Nat. Acad. Sci. USA* 79 (1982), 2554–58.

[36] D. d'Humières, P. Lallemand, Lattice gas cellular automata, a new experimental tool for hydrodynamics, *Physica*, 140A (1986), 326

[37] K. Hwang and F. Briggs, *Parallel Processing and Computer Architecture*, McGraw-Hill, New York (1984)

[38] S. A. Kauffman, Behaviour of randomly constructed genetic nets, in *Towards a Theoretical Biology*, C. H. Waddington (ed.), Vol 3, Edinburgh University Press (1970), 18–46

[39] S. A. Kauffman, Boolean systems, adaptive automata, evolution, in [7], 339–60

[40] K. H. Kim and F. W. Roush, Realizing all linear transformations, *Linear Algebra and Appl.*, 37 (1981), 97–101

[41] S. Kirkpatrick, Models of disordered materials, in *Ill Condensed Matter*, Les Houches, École d'Été de Physique Théorique, session xxi, North-Holland (1979)

[42] S. Kirkpatrick, C. D. Gelatt and M. P. Vecchi, Optimization by simulated annealing, *Science*, 220, 4598 (1983), 671–80

[43] T. Kohonen, *Self-Organization and Associative Memory*, Springer Series in information Sciences, vol. 8, Springer Verlag, Berlin, Heidelberg, New York (1984)

[44] S. R. Kosaraju, On some open problems in the theory of cellular automata, *IEEE Trans. Comp.* 23 (1974), 561–71

[45] A. V. Kulkarni and D. W. L. Yen, Systolic processing and an implementation for signal and image processing, *IEEE Trans. Comp.*, 31, 10 (1982), 1000–9

[46] H. T. Kung, The structure of parallel algorithms, *Adv. Comp.*, 19 (1980), 65–112

[47] H. T. Kung, Why systolic architectures?, *Computer*, 15, 1 (1982), 37–46

[48] H. T. Kung, Programmable systolic chip, *NATO Advanced Study Institute on Microarchitecture of VLSI Computers*, Sogesta, Italy, July 9–20 (1984)

[49] H. T. Kung and C. E. Leiserson, Systolic arrays for VLSI, *Proc. of the Symposium on Sparse Matrices Computations*, I. S. Duff *et al.* (eds), Knoxville, Tenn. (1978), 256–82

[50] S. Y. Kung, VLSI array processors, *IEEE ASSP Magazine*, 2, 3 (1985), 4–22

[51] S. Y. Kung, H. J. Whitehouse and T. Kailath (eds), *VLSI and Modern Signal Processing*, Prentice-Hall, Englewood Cliffs, NJ (1985)

[52] Y. Le Cun, Learning process in an asymmetric threshold network, in [7], 233–40

[53] C. E. Leiserson and J. B. Saxe, Optimizing synchronous systems, *Proc. 22nd Symposium on Foundations of Computer Science* (1981), 23–36

[54] C. E. Leiserson, Area-efficient VLSI computation, PhD Thesis, Carnegie Mellon University, Pittsburgh, Penn. (1981)

[55] W. S. McCulloch and W. Pitts, A logical calculus of the ideas immanent in nervous activity, *Bull. Math. Biophys.*, 5 (1943), 115–33

[56] C. von der Malsburg and E. Bienenstock, Statistical coding and short-term synaptic plasticity: a scheme for knowledge representation in the brain, in [7], 247–72

[57] C. A. Mead and M. A. Conway, *Introduction to VLSI Systems*, Addison-Wesley, Reading, Mass. (1980)

[58] M. Minsky and S. Papert, *Perceptrons, an Introduction to Computational Geometry*, MIT Press, Cambridge, Mass. (1969)

[59] W. L. Miranker and A. Winkler, Spacetime representations of systolic computational structures, IBM Research Report RC9775 (1982)

[60] S. Miyake and K. Fukushima, A neural network model for the mechanism of feature extraction, *Biol. Cybern.*, 50 (1984), 377–84

[61] D. I. Moldovan, On the design of algorithms for VLSI systolic arrays, *Proc. IEEE*, 71, 1 (1983), 113–20

[62] C. Moraga, Systolic Algorithms, Technical Report, Computer Science Department, University of Dortmund, FRG (1984)

[63] C. Moraga, On a case of symbiosis between systolic arrays, *Integration, the VLSI Journal*, 2 (1984), 243–53

[64] J. von Neumann, *Theory of Self Reproducing Automata*, A. W. Burks (ed.), University of Illinois Press, Urbana (1966)

[65] A. Paz, *Introduction to Probabilistic Automata*, Academic Press, New York (1971)

[66] P. Peretto, Collective properties of neural networks, a statistical physics approach, *Biol. Cyb.*, 50 (1984), 51–62

[67] P. Quinton, The systematic design of systolic arrays, in [17], 347–57

[68] M. O. Rabin, Probabilistic automata, *Inf. & Control*, 6 (1966), 230–48

[69] D. Richardson, Tessellation with local transformation, *J. Comput. & Syst. Sci.*, 6 (1972), 373–88

[70] F. Robert, Basic results for the behaviour of discrete iterations, in [7], 33–47

[71] F. Robert, *Discrete Iterations*, Springer Verlag, Berlin, Heidelberg, New York (1986)

[72] Y. Robert, Algorithmique parallèle, Thesis, Grenoble (1986)

[73] Y. Robert and M. Tchuente, Résolution systolique de systèmes linéaires denses, *RAIRO Modélisation et Analyse Numérique*, 19, 2 (1985), 315–26

[74] Y. Robert and D. Trystram, Un réseau systolique orthogonal pour le problème du chemin algébrique, *C. R. Acad. Sci.*, 302, 1, 6 (1986), 241–44

[75] I. G. Rosenberg, The multi-faceted completeness problem of the structural theory of automata, in [17], 375–93

[76] F. Rosenblatt, *Principles of Neurodynamics*, Spartan, 1962

[77] G. Rote, A systolic array algorithm for the algebraic path problem (shortest paths; matrix inversion), *Computing*, 34 (1985), 191–219

[78] D. E. Rumelhart and J. L. McClelland (eds), *Parallel and Distributed Processing: Explorations in the Microstructure of Cognition*, MIT Press, Cambridge, Mass. (1986)

[79] R. Shingai, Maximum period of 2-dimensional uniform neural networks, *Inf. & Control*, 11 (1979), 324–41

[80] A. R. Smith III, Cellular automata theory, Technical Report 2 (1969), Stanford University

[81] M. Tchuente, Parallel calculation of a linear mapping on a computer network, *Linear Algebra & Appl.*, 28 (1979), 223–47

[82] M. Tchuente, Contribution à l'étude des méthodes de calcul pour des systèmes de type coopératif, Thesis, Grenoble (1982)

[83] M. Tchuente, Dynamics and self-organization in one-dimensional arrays, in [7], 21–32

[84] R. Thomas (ed.), *Kinetic Logic*, Lecture Notes in Biomathematics, vol 29, Springer Verlag, Berlin, Heidelberg, New York (1979)

[85] T. Toffoli, CAM: a high-performance cellular automaton machine, in [20], 195–204

[86] J. D. Ullman, Computational aspects of VLSI, Chapter 5: Systolic algorithms, Computer Science Press, Rockville, Md. (1984)

[87] G. Y. Vichniac, Cellular automata models of disorder and organization, in [7], 3–21 and Simulating Physics with cellular automata, *Physica* 10D (1984), 96–116

[88] G. Weisbuch, Un modèle de l'évolution des espèces à trois niveaux, basé sur les propriétés globales des réseaux booléens, *C. R. Acad. Sci.*, 298 (1984), 375–8

[89] G. Weisbuch and F. Fogelman Soulié, Scaling laws for the attractors of Hopfield networks, *J. Phys. Lett.*, 46 (1985), 623–30

[90] U. Weiser and A. Davis, A wavefront notation tool for VLSI design, in *VLSI Systems and Computations*, H. T. Kung *et al.* (eds), Computer Science Press, Rockville, Md. (1981), 226–34

[91] S. Wolfram, Statistical mechanics of cellular automata, *Rev. Mod. Phys.*, 55, 3, (1983), 601–42.

Part 1
Mathematical tools

1 *François Robert*

An introduction to discrete iterations

1.1 Introduction

Discrete dynamical systems can be described, in short, as *iterative models* where *space* and *time* are *discrete*; where the elements involved in the system, called *cells*, are *finite automata*; and where the *interactions* between these elements are moreover *purely local*. As is now well known, these systems go back to J. von Neumann, in the early 1950s [20]. They have received much attention from specialists in the past two decades, and this interest is now widening strongly to the scientific community.

The reason for this is probably that these systems offer a new way for modelling various phenomena. Discrete systems, indeed, have been used with some success in various domains of science; in physics, for example, for spin glass problems, see [15, 18]; and quite recently for discrete simulations in fluid dynamics and in turbulence theory [12], [16]; in chemistry, for diffusion reactions [14]; in biology, mainly for modelling neural networks and genetic nets [15, 17, 21, 26, 28], but also for morphogenesis [9]; in computer science, for pattern recognition, associative memories [11, 15], cellular automata [3, 4, 5, 20, 27, 29], cellular arrays for parallel and systolic computations [19, 22, 25, 27], and so on: see especially recent references [2, 7, 8, 9, 10] on the subject.

These discrete dynamical systems can be presented now as an interesting alternative to classical 'continuous' models using differential systems – both ordinary (ODEs) and partial (PDEs).

First, discrete dynamical systems fit much better than differential systems can to the (essentially) finite discrete digital computing devices of everyday use. When running a discrete model on a computer, there is only a fast computation on bits, and there is no need for a discretisation of the model, nor for a floating-point representation of the reals; consequently, there is no approximation, no rounding errors to consider, but simply an *exact simulation of the model*. Moreover, various phenomena which are in essence discrete, as the real world is, have been recently directly modelled in terms of discrete dynamical systems (see the examples given above). Finally, discrete models are naturally *parallel*, and this fits well with the present development of parallel computing architectures.

A wide field of continuous mathematics has been developed for the last two centuries in analysing classical problems of physics and mechanics. Concerning discrete dynamical systems, on the other hand, theory is still in its infancy. Many important *simulations* have been performed [9, 10, 11, 14, 15, 16, 17, 18, 29] – see also [3, 5]; they provide interesting facts. Many various *theoretical* approaches, also, have been tried [2, 4, 6, 10, 11, 13, 14]. One can easily understand that *there is a need for new mathematical tools*, allowing some coherent analysis of discrete dynamical systems.

Various recent results such as [6, 9, 10, 14, 21, 27] show interesting progress in different directions, for example when a notion of *energy* can be attached to the system, by analogy with classical physics (see [11, 13, 15]). However, there is not yet even the beginning of a unified theory: this field of research is still wide open, and very interesting indeed.

Along these lines, the present chapter offers a way for analysing *discrete iterations*.

In particular, we shall see how well known behavioural results from the domain of *continuous* iterations can be shifted to the context of *discrete* ones. We shall make use mainly of a *discrete metric tool* (a boolean vectorial distance) coupled with a notion of *discrete derivative* (analogous to a Jacobian matrix).

These tools allow basic results to be established for discrete iterations. Here we will examine only some of them (see [24] for an extended analysis). These results, in turn, have been applied in different contexts: in *boolean networks*, where the discrete derivative can be used for studying forcing notions (see [11]) and in *threshold automata*, where attractive fixed points are searched for in designing associative memories [11].

1.2   Basic notions

In what follows, we always consider the *n*-cube $C_n = \{0, 1\}^n$ as our basic space. In many examples, C_n is the set of all possible *configurations* of zeros and ones (or $\cdot$ and $*$) at the *nodes* (often called cells) of a (k, k) grid (with $n = k^2$): see Fig. 1.1.

Of course, the cardinality of C_n is $2^n = 2^{k*k}$, which is huge, even for small k ($k = 5 : 2^{25} = 33\,554\,432$; $k = 10 : 2^{100} \approx 10^{30}$). The examples given below are much smaller ($n = 3$ or 4).

An element x of C_n is then an *n*-tuple $x = (x_1, \ldots, x_n)$ of zeros and ones. Given such an x, the *immediate neighbourhood* (or first neighbourhood) of x, denoted by $V_1(x)$, is defined as

$$V_1(x) = \{x, \tilde{x}^1, \ldots, \tilde{x}^n\}$$

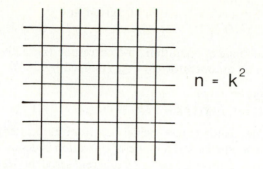

Figure 1.1 The *n*-cube as a $k \times k$ grid

where $\tilde{x}^i = (x_1, \ldots, \bar{x}_i, \ldots, x_n)$ is a *first neighbour* of x ($i = 1, 2, \ldots, n$), with the convention that $\bar{0} = 1$, and $\bar{1} = 0$.

Notice that, given a configuration x at the nodes of the grid, a first neighbour of x is nothing else than x itself where at most one cell has been changed ($0 \mapsto 1$ or $1 \mapsto 0$).

Example
$k = 3, n = 9$

*	○	○		*	○	○

A configuration x A first neighbour of x

Of course, it is possible to define wider neighbourhoods, such as, for example, the second, the third, ... or the kth neighbourhood of an element x of C_n.

Now, we will consider a given, fixed mapping, say F, from C_n into itself. The problem is (remember that card(C_n) is huge) to say something about the behaviour of the following discrete iteration in C_n:

$$\begin{cases} x^0 \quad \text{given in } C^n. \\ x_i^{r+1} = F(x^r) \quad (r = 0, 1, \ldots) \end{cases}$$

that we detail in

$$\left.\begin{array}{l} x^0 = (x_1{}^0, x_2{}^0, \ldots, x_n{}^0) \quad \text{given in } C_n. \\ x^{r+1} = f_i(x_1{}^r, x_2{}^r, \ldots, x_n{}^r) \quad (i = 1, 2, \ldots, n). \end{array}\right\} \quad (1.1)$$

For obvious reasons, the functions f_i (from C_n into $C_1 = \{0, 1\}$) are sometimes referred to as the *local transition functions*, whereas F is known as the *global transition function*.

Equations (1.1), in fact, are an elementary model for the *parallel processing* of an automata network defined by the given operator F. A *serial mode* of operation for this automaton would be

$$y^0 = (y_1^0, y_2^0, \ldots, y_n^0) \quad \text{given in } C_n.$$
$$y_i^{r+1} = f_i(y_1^{r+1}, \ldots, y_{i-1}^{r+1}, y_i^r, \ldots, y_n^r) \quad (i = 1, 2, \ldots, n). \qquad (1.2)$$

This serial mode of operation on F corresponds exactly to a parallel mode of operation on an operator $G = (g_i)$ defined by

$$g_1(y) = f_1(y)$$
$$g_i(y) = f_i(g_1(y), \ldots, g_{i-1}(y), y_i, \ldots, y_n) \quad (i = 1, 2, \ldots, n).$$

Here, G stands for Gauss–Seidel (operator associated to F).

Of course, such a serial mode of operation is defined up to a permutation of the indices i (there are $(n - 1)!$ different serial modes for one parallel mode of operation). More general kinds of serial, or serial-parallel, or even chaotic (random) modes of operation can be also considered (see [11, 23]).

As C_n, though large, is finite, all these iterative processes finally end up as a *cycle*, or even a *fixed point* (that is, a stable configuration), generally after some *transitory phase*. The problem is now to say something interesting about these cycles, fixed points, and the *transitory length*.

Notice that the fixed points remain unchanged when passing from one mode of operation to another: they are simply the fixed points ($v = F(v)$) of the given operator F on C_n. However, the *cycles generally differ* from one mode of operation to another (see Example 1 below).

In classical examples, the local transition functions we consider are actually local, that is each f_i depends only on some variables x_j. Typical examples are the well-known von Neumann and Moore neighbourhoods on the (k, k) grid, as shown in Fig. 1.2.

(a) Von Neumann
  f_i depends only on x_{i-k},
  $x_{i-1}, x_i, x_{i+1}, x_{i+k}$

(b) Moore
  f_i depends only on x_{i-k-1},
  $x_{i-k}, x_{i-k+1}, x_{i-1}, x_i, x_{i+1}$,
  $x_{i+k-1}, x_{i+k}, x_{i+k+1}$

Figure 1.2 Definition of neighbourhoods:

Indeed, the dependence of the f_i on the x_j can be visualised equivalently either by a *connectivity graph* (of n vertices P_i), where an arc joins P_j to P_i if f_i really depends on x_j.

or by an *incidence matrix*: that is an $(n \times n)$ boolean matrix $B(F) = (b_{ij})$

with $\begin{cases} b_{ij} = 1 \text{ if there is an arc from } P_j \text{ to } P_i \\ b_{ij} = 0 \text{ otherwise.} \end{cases}$

Notice that the connectivity graph for F has n vertices, whereas the iteration graph for F (that is the graph drawn on C_n by joining any x to $F(x)$) has 2^n vertices.

Example 1 $n = 3$, with

$$F(x) = \begin{cases} f_1(x_1, x_2, x_3) = x_1 + x_2 \cdot \overline{x_3} \\ f_2(x_1, -, x_3) = x_1 \cdot \overline{x_3} \\ f_3(-, x_2, -) = x_2 \end{cases} \quad \text{(boolean notations)}$$

The incidence matrix for F is $B(F)$ and its connectivity graph is as in Fig. 1.3, where:

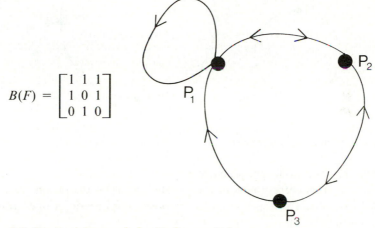

$$B(F) = \begin{bmatrix} 1 & 1 & 1 \\ 1 & 0 & 1 \\ 0 & 1 & 0 \end{bmatrix}$$

Figure 1.3 Connectivity graph for F of example 1

The tabulation for F and its iteration graph are as in Fig. 1.4.

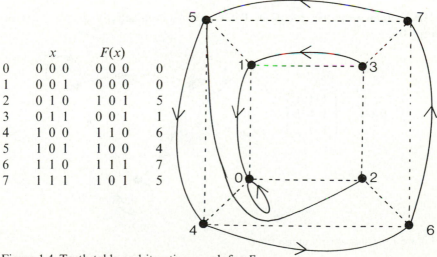

	x	$F(x)$	
0	0 0 0	0 0 0	0
1	0 0 1	0 0 0	0
2	0 1 0	1 0 1	5
3	0 1 1	0 0 1	1
4	1 0 0	1 1 0	6
5	1 0 1	1 0 0	4
6	1 1 0	1 1 1	7
7	1 1 1	1 0 1	5

Figure 1.4 Truth table and iteration graph for F

This iteration graph can be redrawn as in Fig. 1.5.

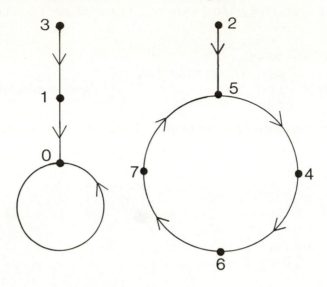

Figure 1.5 Iteration graph for F

There is one fixed point and one cycle.

Of course, this corresponds to the parallel mode of operation of F. The Gauss–Seidel operator (associated to the basic serial mode of operation on F) is then defined by

$$G(x) = \begin{cases} g_1(x_1, x_2, x_3) = f_1(x_1, x_2, x_3) = x_1 + x_2 \cdot \overline{x_3} \\ g_2(x_1, x_2, x_3) = (x_1 + x_2 \cdot \overline{x_3}) \cdot \overline{x_3} = (x_1 + x_2) \cdot \overline{x_3} \\ g_3(x_1, x_2, x_3) = (x_1 + x_2) \cdot \overline{x_3} \end{cases}$$

The incidence matrix for G is $B(G)$ and its connectivity graph is as in Fig. 1.6.

$$B(G) = \begin{bmatrix} 1 & 1 & 1 \\ 1 & 1 & 1 \\ 1 & 1 & 1 \end{bmatrix}$$

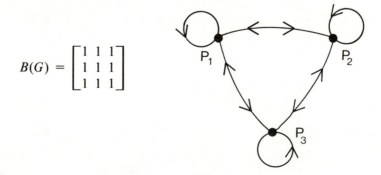

Figure 1.6 Connectivity graph for G

The tabulation for G follows and its iteration graph is as in Fig. 1.7.

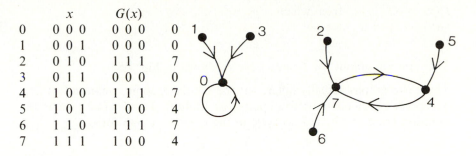

	x	$G(x)$	
0	0 0 0	0 0 0	0
1	0 0 1	0 0 0	0
2	0 1 0	1 1 1	7
3	0 1 1	0 0 0	0
4	1 0 0	1 1 1	7
5	1 0 1	1 0 0	4
6	1 1 0	1 1 1	7
7	1 1 1	1 0 0	4

Figure 1.7 Iteration graph for G

We verify in passing that G does have the same fixed point as F; however, G has not the same cycle as F.

In the general case, the problem of comparing the shapes of the iteration graphs for F and G is an interesting, but difficult question (see [24]).

Here is another example:

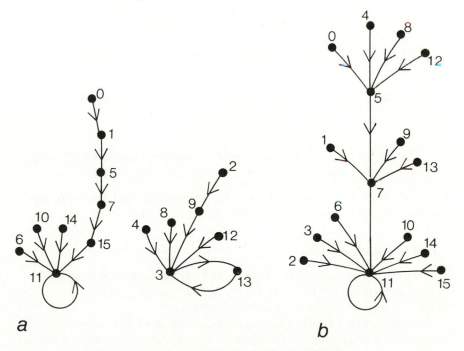

a

b

Figure 1.8 (a) Iteration graph for F; (b) Iteration graph for G

Example 2 $n = 4$ (boolean notations)

$$f_1(x) = x_3 \qquad\qquad\qquad g_1(x) = x_3$$
$$f_2(x) = \overline{x_1} \cdot x_4 \quad \text{from which} \quad g_2(x) = \overline{x_3} \cdot x_4$$
$$f_3(x) = x_1 + x_2 \quad \text{we get:} \quad g_3(x) = x_3 + \overline{x_3} \cdot x_4$$
$$f_4(x) = 1 \qquad\qquad\qquad g_4(x) = 1$$

The iteration graphs for F and G are as shown in Fig. 1.8.

These two graphs actually differ, but they do have some common features. The point 11 is a common fixed point for F and G. There is one cycle in the iteration graph for F, and no cycle in the iteration graph for G.

1.3   A boolean vectorial distance on C_n

We will make use of the following boolean vector distance d on C_n:

> if $x = (x_1, \ldots, x_n)$ and $y = (y_1, \ldots, y_n)$
> then $d(x, y)$ is the n-vector $[d_i(x_i, y_i); 1 \le i \le n]'$
> where $d_i(x_i, y_i) = 0$ if $x_i = y_i$, 1 otherwise.

For example,

$$d \begin{bmatrix} 0 & 0 \\ 0 & 1 \\ 1 & 0 \\ 1 & 1 \end{bmatrix} = \begin{bmatrix} 0 \\ 1 \\ 1 \\ 0 \end{bmatrix}$$

Notice that $d(x, y)$ is to be considered as a *boolean vector* and that we get the usual axioms for the distance d:

$$d(x, y) = 0 \text{ iff } x = y$$
$$d(x, y) = d(y, x)$$
$$d(x, y) + d(y, z) \ge d(x, z)$$
$$\qquad\quad \uparrow \qquad\qquad \uparrow$$

Boolean sum   componentwise in $\{0, 1\}^n = C_n$
$(1 + 1 = 1)$ $(0 \le 0 \le 1 \le 1)$

Equipped with this topological tool, C_n is complete, and the topology is discrete; the converging sequences are the finally stationary sequences.

1.4   The discrete derivative

The *discrete derivative* of F at a point x in C_n is the boolean matrix $F'(x) = (f_{ij}(x))$ defined by $f_{ij}(x) = 1$ if $f_i(x) \ne f_i(\tilde{x}^j)$ [that is if $f_i(x_1, \ldots, x_n) \ne f_i(x_1, \ldots, \overline{x}_j, \ldots, x_n)$] and $f_{ij}(x) = 0$ otherwise.

This derivative will play the role of the *Jacobian* of F. Computing $F'(x)$ needs $n^2 + n$ evaluations of functions f_i. For instance, in example 1 above, we get

$$F'(0) = \begin{bmatrix} 1 & 1 & 0 \\ 1 & 0 & 0 \\ 0 & 1 & 0 \end{bmatrix} \text{ and } F'(7) = \begin{bmatrix} 1 & 0 & 0 \\ 0 & 0 & 1 \\ 0 & 1 & 0 \end{bmatrix}$$

There are some basic and interesting relations between the discrete derivative $F'(x)$, the boolean vector distance d and the incidence matrix $B(F)$ of F:

Proposition 1
See reference [24].

(i) $\forall x \in C_n, F'(x) \le B(F)$

(ii) $\sup_{x \in C_n} \{F'(x)\} = B(F)$

(iii) $\forall x \in C_n, \forall j, d(F(x), F(\tilde{x}^j)) = F'(x) \cdot \underbrace{d(x, \tilde{x}^j)}_{\substack{e_j}}$

$$\underbrace{\phantom{d(x, \tilde{x}^j)}}_{j\text{th column of } F'(x)}$$

Proof   To prove (i), it is sufficient to show that if $B(F)$ has a zero in position i, j, then the same is true for $F'(x)$ independent of x. Clearly, whenever f_i is by hypothesis independent of the jth variable x_j, then we have necessarily for all $x \in C_n$ that $f_i(x) = f_i(\tilde{x}^j)$. Since this means that $f_{ij}(x) = 0$, we have proved (i).

Furthermore, it follows from (i) that

$$\sup_{x \in C_n} \{F'(x)\} \le B(F)$$

(The sup and $\le$ are taken pointwise on the boolean matrices of size n, n.)

In order to show that the above inequality is indeed an equality, it is sufficient to prove that if $B(F)$ has a 1 in position i, j then there exists an $x \in C_n$ such that $F'(x)$ also has a 1 in position i, j.

Now, saying that f_i actually depends on x_j is equivalent to: $b_{ij} = 1$. Therefore there exists $x \in C_n$ such that evaluating f_i in $x = (x_1, \ldots, x_j, \ldots, x_n)$ and $\tilde{x}^j = (x_1, \ldots, \bar{x}_j, \ldots, x_n)$ results in two different values, that is

$$f_i(x) \ne f_i(\tilde{x}^j)$$

from which $f_{ij}(x) = 1$, and we have proved (ii).

The proof of (iii) is easy: by definition, $d(x, \tilde{x}^j) = e_j$, so that $F'(x) \cdot d(x, x^j)$ represents the jth column of $F'(x)$, which by definition coincides with the vector $d(F(x), F(\tilde{x}^j))$.   □

Point (iii) of Proposition 1 above shows that for any x and one of its n first neighbours $\tilde{x}^j$, the distance $d(F(x), F(\tilde{x}^j))$ can be evaluated exactly. When y is no longer a first neighbour of x, $d(F(x), F(y))$ can only be bounded:

Proposition 2
See reference [24].

$$\forall x, \, y \in C_n, \, d(F(x), F(y)) \leq \sup_{z \in [x, y]} \{F'(z)\} \cdot d(x, \, y)$$

for an (arbitrary) chain $[x, \, y]$ of minimal length on the n-cube C_n.

The proof is somewhat technical, and the reader is referred to [24]. Moreover we get a basic inequality:

Proposition 3
See reference [24].

$$\forall x, \, y \in C_n, \, d(F(x), F(y)) \leq B(F) \cdot d(x, \, y)$$

which can be either obtained from Proposition 2 and Proposition 1 (ii), or proved directly.

1.5   Application: local convergence in the first neighbourhood of a fixed point (cf [24]).

Let $\S = F(\S)$ be a *fixed point* of F in C_n (that is a stable configuration for F). We say that $\S$ is *attractive* (in its first neighbourhood $V_1(\S)$) if

(i) $F(V_1(\S)) \subseteq V_1(\S)$
(ii) for any $x^0 \in V_1(\S)$, the iteration $x^{r+1} = F(x^r)$ remains in $V_1(\S)$, according to (i), and reaches $\S$ in at most n steps. Then $x^n = F^n(x^0) = \S$.

This notion of attraction can be characterised by using the tools introduced above:

Proposition 4   A necessary and sufficient condition for $\S$ to be attractive in its first neighbourhood is the following:

(i) $F'(\S)$ has at most one 1 per column.
(ii) The *boolean spectral radius* $\rho(F'(\S))$ of $F'(\S)$ is zero; that is, there exists a permutation matrix P such that the matrix $P^tF'(\S)P$ is strictly lower triangular (or, equivalently, there exists an integer $p \leq n$ such that $(F'(\S))^p = 0$ (boolean power).

Proof   (a) The sufficient conditions. If (i) and (ii) are satisfied, then we have for all $x \in V_1(\S)$:

$$d(F(x), \S) = F'(\S) \cdot \underbrace{d(x, \S)}_{e_i}$$

$$\underbrace{}_{i\text{th column of } F'(\S)}$$

By assumption, the columns of $F'(\S)$ may only be zero or a basis vector, from which it follows that $F(x) \in V_1(\S)$. Hence, for all $x^0 \in V_1(\S)$, the sequence $x^{r+1} = F(x^r)$ remains in $V_1(\S)$, and we have

$$d(x^r, \S) = d(F(x^{r-1}), \S) = F'(\S) \cdot d(x^{r-1}, \S) = (F'(\S))^r \cdot d(x^0, \S)$$

Since $\rho(F'(\S)) = 0$, there exists an integer $p \leq n$ such that $(F'(\S))^p = 0$ (boolean power). Therefore, $d(x^p, \S) = 0$, and $x^p = \S$. In other words the sequence reaches the fixed point $\S$ in at most p steps.

(b) The necessary conditions. Suppose that $\S$ is attractive in $V_1(\S)$ and that one of the columns of $F'(\S)$, say the jth, contains more than one 1. We then have that

$$d(F(\tilde{\S}^j), \S) = F'(\S) \cdot \underbrace{d(\tilde{\S}^j, \S)}_{e_j}$$

$$\underbrace{\qquad\qquad\qquad}_{\text{the } j\text{th column of } F'(\S)}$$

This shows that $F(\tilde{\S}^j)$ is not a first neighbour of $\S$, since it differs from $\S$ in more than one component. Therefore F does not map $V_1(\S)$ onto itself, which is a contradiction with (i).

We now show that the boolean spectral radius of $F'(\S)$ is necessarily zero. From the assumptions, it follows that $Ax \in V_1(\S)$, $F^n(x) = \S$. In particular for $x = \tilde{\S}^j$, we get $d(F(\tilde{\S}^j), \S) = F'(\S) \cdot e_j$

From the assumptions, $F(\tilde{\S}^j)$ is also in $V_1(\S)$, from which

$$d(F^2(\tilde{\S}^j), \S) = F'(\S) \cdot d(F(\tilde{\S}^j), \S) = (F'(\S))^2 \cdot \underbrace{d(\tilde{\S}^j, \S)}_{e_j}$$

and by recurrence

$$d(F^n(\tilde{\S}^j), \S) = (F'(\S))^n \cdot e_j \ (j = 1, 2, \ldots, n).$$

From the assumptions, the left-hand side of the equation is zero. Thus all columns of $(F'(\S))^n$ are zero. Then $(F'(\S))^n$ (boolean power) is the zero matrix, from which it results that the boolean spectral radius $\rho(F'(\S))$ of $F'(\S)$ is zero. □

Example 3　$n = 3; \S = (0 \ 1 \ 1)$ and

$$F'(\S) = \begin{bmatrix} 0 & 0 & 0 \\ 1 & 0 & 1 \\ 0 & 0 & 0 \end{bmatrix}$$

$F'(\S)$ satisfies the required conditions. One easily checks that necessarily:

	x	$F(x)$		
	0	0 0 0		
$\tilde{s}^2$	1	0 0 1	0 1 1 3	s
$\tilde{s}^3$	2	0 1 0	0 0 1 1	$\tilde{s}^2$
$\tilde{s}$	3	0 1 1	0 1 1 3	s
	4	1 0 0		
	5	1 0 1		
	6	1 1 0		
$\tilde{s}^1$	7	1 1 1	0 0 1 1	$\tilde{s}^2$

This shows that the iteration graph must contain that shown in Fig. 1.9.

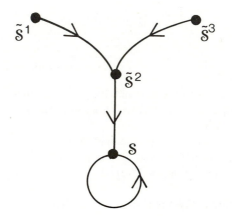

Figure 1.9

Indeed s is attractive in its first neighbourhood $V_1(s)$.

    This notion of attractive fixed point is meaningful in the context of automata networks. In fact, it means that a stable configuration s would attract its first neighbours in at most n steps, by passing only through first neighbours of s in the meantime.

1.6   Massive neighbourhoods

It is possible to define attraction in wider subsets than the first neighbourhood of a fixed point s of F. Characterisations of these notions quickly become cumbersome. However one can give simple sufficient conditions of attraction in a massive neighbourhood of a fixed point.

Definition   A *massive* neighbourhood V of an element $x \in C_n$ is a subset of C_n containing x, and such that:

    If $u \in V$, then every $y \in C_n$ such that $d(x, y) \le d(x, u)$ is also in V.

An example is shown in Fig. 1.10.

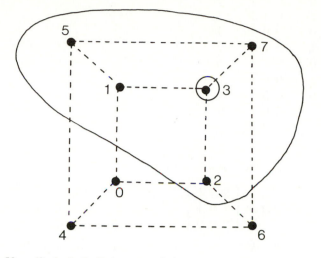

Figure 1.10 V $=$ {3, 1, 5, 7, 2} is a massive neighbourhood of 3. The associated massive neighbourhood of 0 is: W $=$ {0, 2, 6, 4, 1}.

Definition   A fixed point § $=$ $F(\S)$ of F is said to be attractive in a massive neighbourhood V of § if

(i) $F(V) \subseteq V$
(ii) For any $x^0 \in V$, the iteration $x^{r+1} = F(x^r)$ (which remains in V from (i)) reaches § in at most n steps.

Below are sufficient conditions for a fixed point § to be attractive in a given massive neighbourhood V of §

Proposition 5   Let V be a massive neighbourhood of a fixed point § of F and let W $=$ {$d(\S, x)$; $x \in V$} be the associated (massive) neighbourhood of 0 in C_n. Furthermore, let

$$M = \sup_{z \in V}\{F'(z)\}$$

If we have $M \cdot W \subseteq W$ and $\rho(M) = 0$ (boolean spectral radius), then § is attractive in V.

Proof
The reader is referred to [24] for a proof.

Example 4 $n = 3$; § $= (0, 1, 1) = 3$; V $=$ {3, 1, 5, 7, 2} is the massive neighbourhood of 3 considered just above; W $=$ {0, 2, 6, 4, 1} is the associated massive neighbourhood of 0. Consider the following F:

	x	$F(x)$	
0	0 0 0	0 0 1	1
1	0 0 1	0 1 1	3
2	0 1 0	0 0 1	1
3	0 1 1	0 1 1	3
4	1 0 0	1 0 1	5
5	1 0 1	0 0 1	1
6	1 1 0	0 0 1	1
7	1 1 1	0 0 1	1

We get

(i) $F(3) = 3 : 3$ is a fixed point for F.

(ii) $M = \sup\limits_{z \in V}\{F'(z) = \begin{bmatrix} 0 & 0 & 1 \\ 1 & 0 & 1 \\ 0 & 0 & 0 \end{bmatrix}$

We now have $M \cdot W = \{0, 2, 6\} \subseteq W$ and $\rho(M) = 0$. We may easily verify the local convergence towards 3 in the massive neighbourhood $V = \{3, 1, 5, 7, 2\}$ of 3 since the iteration graph of F is as shown in Fig. 1.11.

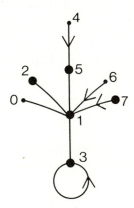

Figure 1.11 Example 4: iteration graph of F (black dots denote the elements of V)

1.7   Attractive cycles (cf [24])

Let $\{\S_1, \S_2, \ldots, \S_p\}$ be a *cycle of length p* for F, that is

$$F(\S_1) = \S_2, \; F(\S_2) = \S_3, \; \ldots, \; F(\S_p) = \S_1$$

with the $\S_i$ all different.

The notion of attractive fixed point can be extended to a cycle; we only quote a basic result:

Proposition 6 A necessary and sufficient condition for the cycle $\{\S_1, \ldots, \S_p\}$ of F to be attractive in its first neighbourhood is:

(i) $F'(\S_i)$ has at most one 1 per column ($i = 1, 2, \ldots, p$)
(ii) the boolean spectral radius of the matrix (boolean product) $F'(\S_p) \ldots F'(\S_1)$ is equal to zero

Here is an example:

Example 5 $n = 3$; F is defined by its table and its iteration graph (Fig. 1.12).

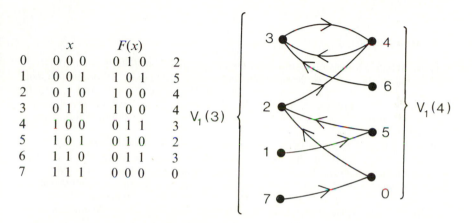

	x	$F(x)$	
0	0 0 0	0 1 0	2
1	0 0 1	1 0 1	5
2	0 1 0	1 0 0	4
3	0 1 1	1 0 0	4
4	1 0 0	0 1 1	3
5	1 0 1	0 1 0	2
6	1 1 0	0 1 1	3
7	1 1 1	0 0 0	0

$V_1(3)$ $\quad$ $V_1(4)$

Figure 1.12 Example 5: iteration graph of F

F admits an attractive cycle of length 2:$\{3, 4\}$. Indeed one verifies that

$$F'(3) = \begin{bmatrix} 1 & 0 & 0 \\ 0 & 0 & 0 \\ 0 & 1 & 0 \end{bmatrix} \text{ and } F'(4) = \begin{bmatrix} 0 & 0 & 0 \\ 0 & 0 & 0 \\ 1 & 0 & 1 \end{bmatrix}$$

both have at most one 1 per column, and that

$$F'(3) \cdot F'(4) = \begin{bmatrix} 0 & 0 & 0 \\ 0 & 0 & 0 \\ 0 & 0 & 0 \end{bmatrix} \text{ and } F'(4) \cdot F'(3) = \begin{bmatrix} 0 & 0 & 0 \\ 0 & 0 & 0 \\ 1 & 1 & 0 \end{bmatrix}$$

both have a boolean spectral radius zero.

1.8 Conclusions

We have tried to show in this introductory chapter how it is possible to study elementary but basic behaviours of discrete iterations, with tools such as a vectorial boolean distance and the use of a discrete derivative (more can be said: see [24] for a contraction theory, for comparisons between different iteration graphs, for a Newton method in C_n, and so on).

Of course, it is not claimed here that these tools are the only ones to be used. Other tools, such as invariants, or energy/entropy notions, for example, have proved to be powerful (see [11, 13] and also, more generally: [6, 7, 8, 9, 10, 15, 27, 29] and the bibliography of [24]).

In conclusion, it should simply be emphasised that iterative properties of discrete dynamical systems and automata networks are an interesting research area, where new tools and new ways of understanding still need to be elaborated.

1.9 References

[1] A. Arbib, *Theories of abstract Automata*, Prentice-Hall (1969)

[2] E. Bienenstock, F. Fogelman Soulié and G. Weisbuch (eds), *Disordered Systems and Biological Organization*, NATO ASI Series in Systems and Computer Science, F20, Springer Verlag, Berlin, Heidelberg, New York (1986)

[3] A. W. Burks, *Essays on Cellular Automata*, University of Illinois Press, Urbana (1970)

[4] E. F. Codd, *Cellular Automata*, Academic Press, New York (1968)

[5] J. H. Conway, E. R. Berlekampf and R. K. Guy, *Winning Ways for your Mathematical Plays*, Vol. 2, Chap. 25, Academic Press, New York (1982)

[6] M. Cosnard and E. Goles, Dynamique d'un automate à mémoire modélisant le fonctionnement d'un neurone, *C. R. Acad. Sci.*, 299, 1, 10 (1984), 459–61

[7] M. Cosnard, J. Demongeot and A. Lebreton (eds), *Rhythms in Biology and other Fields of Applications*, Lecture notes in Mathematics 49, Springer Verlag, Berlin, Heidelberg, New York (1983)

[8] J. Della Dora, J. Demongeot and B. Lacolle, (eds), *Numerical Methods in the Study of Critical Phenomena*, Springer Verlag, Berlin, Heidelberg, New York (1981)

[9] J. Demongeot, E. Goles and M. Tchuente (eds), *Dynamical Systems and Cellular Automata*, Academic Press, New York (1985)

[10] D. Farmer, T. Toffoli and S. Wolfram (eds), *Cellular Automata*, North-Holland, Amsterdam (1984)

[11] F. Fogelman Soulié, Contributions à une théorie du calcul sur réseaux, Thesis, Grenoble (1985)

[12] U. Frisch, B. Hasslacher and Y. Pomeau, A lattice gas automaton for the Navier–Stokes equation, *Phys. Rev. lett.* (1986)

[13] E. Goles: Comportement dynamique de réseaux d'automates, Thesis, Grenoble (1985)

[14] J. M. Greenberg, B. D. Hassard and S. P. Hastings, Pattern formation and periodic structures in systems modelled by reaction diffusion equations, *Bull. Am. Math. Soc.*, 84, 6 (1978), 1296–1327

[15] J. J. Hopfield, Neural networks and physical systems with emergent collective computational abilities, *Proc. Nat. Acad. Sci. USA*, 79 (1982), 2554–8

[16] D. d'Humières, P. Lallemand, Lattice gas cellular automata, a new experimental tool for hydrodynamics, *Physica*, 140A (1986), 326

[17] S. Kauffman, Behaviour of randomly constructed genetic nets, in *Towards a Theoretical Biology*, Vol 3, Edinburgh University Press (1970), 18–46

[18] S. Kirkpatrick, Models of disordered materials, in *Ill Condensed Matter*, Les Houches, École d'Été de Physique Théorique, North Holland (1979)

[19] C. A. Mead and M. A. Conway, *Introduction to VLSI Systems*, Addison-Wesley Reading, Mass. (1980)

[20] J. von Neumann, *Theory of Self Reproducing Automata*, A. W. Burks (ed.), University of Illinois Press, Urbana (1966)

[21] P. Peretto, Collective properties of neural networks; a statistical physics approach, *Biol. Cyb.*, 50 (1984), 51–62

[22] P. Quinton, The systematic design of systolic arrays, in [9]

[23] F. Robert, Basic results for the behaviour of discrete iterations, in [2], 33–47

[24] F. Robert, *Discrete Iterations*, Springer Verlag, Berlin, Heidelberg, New York (1986)

[25] Y. Robert, Algorithmique parallèle, Thesis, Grenoble (1986)

[26] R. Shingai, Maximum Period of 2-dimensional uniform neural networks, *Inf. & Control*, (1979), 324–41

[27] M. Tchuente, Contribution à l'étude des méthodes de calcul pour des systèmes de type coopératif, Thesis, Grenoble (1982)

[28] R. Thomas (ed.), *Kinetic Logic*, Lecture Notes in Biomathematics, Vol 29, Springer Verlag, Berlin, Heidelberg, New York (1979)

[29] S. Wolfram, Statistical mechanics of cellular automata, *Rev. Mod. Phys.*, 55, 3, (1983), 601–42

This chapter is an updated and somewhat extended version of [23], including new examples in particular.

Appendix A.1 *Françoise Fogelman Soulié*

Boolean networks

A.1.1 Introduction

Boolean networks have been studied in various contexts: as a modelling tool in biology [7, 8, 12], in social sciences [5] or as a discrete dynamical system [3, 9, 10, 11]. This appendix gives some results which have been derived to characterise the geometrical structure of the limit cycles in boolean networks.

A.1.2 Definitions

A *boolean mapping* in n variables is a mapping $f\colon \{0, 1\}^n \to \{0, 1\}$.
A *boolean network* of n elements is a mapping $F\colon \{0, 1\}^n \to \{0, 1\}^n$, where $F = (f_1, \ldots, f_n)$, and all f_i are boolean mappings in n variables.

A boolean network can be viewed as a set of interconnected elementary processors (see Fig. A.1.1).

Let $F = (f_1, \ldots, f_n)$ be a boolean network of n elements. If all f_i depend on $k \leq n$ variables at most, then network F is said to be of *connectivity* k. The *connection graph* of F will be denoted $\mathscr{C} = (X, \mathscr{W})$ with $X = \{1, \ldots, n\}$ and $(i, j) \in X \times X$ is in $\mathscr{W}$ iff f_j really depends on variable x_i:

If the n elements of network F lie in some topological space (for example $\mathbb{R}$ or $\mathbb{R}^2$), then we will say that F is *locally connected* or displays local – or short-range – interactions if k is small with respect to n (for example $k \leq 4$), and each element i has its function f_i depending on k *neighbouring* elements. This is for example the case in cellular networks.

Let $F = (f_1, \ldots, f_n)$ be a boolean network of n elements and $x^0 \in \{0, 1\}^n$. A *trajectory* of F associated to iteration F_d, starting at x^0, is a sequence $[x(t)]_{t \geq 0}$ in $\{0, 1\}^n$ such that:

$$\forall t \geq 0, \ x(t + 1) = F_d[x(t)] \quad \text{and} \quad x(0) = x^0$$

where F_d is an *interation mode* on F:

for the parallel iteration, we have $F_d = F$,

for the sequential iteration associated to the permutation π of $\{1, \ldots, n\}$, we have $F_d = F_\pi$

(a)

1	11	10	13	9	5	10	7	12	4	9	12	1	9	12	3
3	7	12	13	10	5	13	8	1	11	9	9	14	7	6	11
14	8	10	4	4	5	6	5	8	4	8	9	8	13	12	2
2	12	12	2	4	9	8	5	11	12	3	1	13	8	14	6
11	14	7	14	12	13	2	11	14	3	6	14	9	4	6	4
12	12	2	10	8	5	14	14	8	4	7	10	2	14	7	4
10	6	13	3	2	13	8	9	13	10	10	8	1	14	4	14
12	13	2	8	2	1	9	4	2	6	7	3	4	8	4	3
13	13	5	11	13	6	10	8	11	6	12	4	8	12	11	5
10	6	13	7	2	9	10	10	4	2	6	5	2	8	10	14
8	11	8	2	4	6	10	4	7	12	13	3	4	3	8	1
4	8	12	9	6	5	6	14	1	12	3	7	13	7	11	1
1	14	2	11	2	4	5	7	3	3	4	5	12	10	11	13
6	9	3	12	14	6	9	12	12	5	11	3	11	11	13	8
12	10	12	8	5	12	1	11	1	13	13	2	6	10	11	11
5	14	7	12	1	9	10	13	12	9	12	3	4	2	7	2

(b)

$$
\begin{array}{ccc}
 & \downarrow\underline{} & \\
\longrightarrow t_2 & \Leftarrow\ \uparrow & \longrightarrow nor \longleftarrow \\
\downarrow & & \downarrow \\
nor \longrightarrow & \Leftrightarrow\ \longleftarrow & \Leftrightarrow \\
\uparrow & \downarrow\longleftrightarrow\ \uparrow & \uparrow \\
\longrightarrow nand \longleftarrow & \Leftarrow & \longrightarrow or \longleftarrow \\
 & \uparrow &
\end{array}
$$

(c)

N^*	Name	Table	v^*	x^*_1	x^*_2
0	*contr.*	00 00	0	0.1	0.1
1	*nor*	10 00	0	1	1
2	$\Rightarrow$	00 10	0	0	1
3	$\bar{t}_2$	10 10	1 0	— —	0 1
4	$\Leftarrow$	01 00	0	1	0

5	$\bar{t}_1$	11 00	0 1	1 0	— —
6	*xor*	01 10	—	—	—
7	*nand*	11 10	1	0	0
8	*and*	00 01	0	0	0
9	⇔	10 01	—	—	—
10	t_1	00 11	0	0	—
11	⇐	10 11	1	1	0
12	t_2	01 01	0 1	— —	0 1
13	⇒	11 01	1	0	1
14	*or*	01 11	1	1	1
15	*taut*	11 11	1	0.1	0.1

Figure A.1.1 Boolean network with connectivity 2 and regular connections between nearest neighbours, drawn on a torus: (a) the transition function f_i for each automaton i in the network; (b) a small part of the network with connections; (c) list of boolean mappings in 2 variables

As the state space is finite, all trajectories are ultimately periodic. It has been shown previously on various examples (see Chapter 1) that parallel and sequential iterations have different limit cycles. However, various observations have been made on simulations of the dynamical behaviour of boolean networks:

the number of limit cycle is small ($n^{1/2}$ if n is the size of the network [7]). most of the states are only transitory, in other words, the limit set of the iteration (i.e. the set of those states which are part of a limit cycle) contains very few states.

in any given limit cycle, most of the automata remain stable or oscillate very rarely.

These observations will allow us to compare the limit cycles.

A.1.3 Stable core

Definition
Let $F: \{0, 1\}^n \to \{0, 1\}^n$ be a boolean network and $C = (s(1), \ldots, s(T))$ be a limit cycle of period T for an iteration F_d of F.
The *stable part* of C is the set

$$S_C = \{i(1, \ldots, n): \forall t(1, \ldots, T), s_i(t) = s_i(1)\}$$

The *oscillating part* of C is the set

$$O_C = \{1, \ldots, n\}/S_C$$

The *stable core* of F for iteration F_d is then:

$$S = \cap_C S_C$$

The *oscillating core* of F for iteration F_d is then:

$$O = \cap_C O_C$$

where the intersection runs on all limit cycles C for iteration F_d (see Fig. A.1.2).

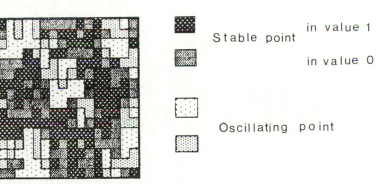

in value 1
Stable point
in value 0

Oscillating point

Figure A.1.2 Limit cycle C for the network of Fig. A.1.1. The stable part S_C is shown in dark and the oscillating part O_C in light.

Of course, it is impossible to compute the cores exactly: this would require running the iteration on all 2^n possible initial conditions. We can only hope to find an approximation of the cores, for example by testing a limited number of initial conditions (Fig. A.1.3).

(a)

```
 106 1000 1000 1000 1000    0    0    0    0    0  211  211  991 1000 1000  220
 106 1000 1000  901 1000    0    0    0    0    0  211  220  220 1000  991  106
1000 1000  901  901 1000 1000 1000    0    0    0    0  220 1000 1000 1000    0
1000 1000  901    0 1000 1000    0    0    0    0    0    0 1000 1000 1000  998
   0    0    0    0    0  823  823    0    0    0    0    0    0    0  998  998
   0    0    0    0    0 1000 1000 1000 1000    0    0    0    0    0  998    0
   0    0    0    0    0    0 1000 1000 1000 1000    0    0    0    0    0    0
   0    0    0    0    0    0 1000    0 1000 1000    0    0    0    0    0    0
1000    0    0    0    0    0    0    0    0 1000 1000    0    0    0    0 1000
   0    0    0    0    0    0    0    0    0 1000 1000    0    0    0 1000 1000
   0    0    0    0    0    0    0    0    0 1000    0    0    0    0 1000 1000
   0    0    0  614  614    0    0    0    0    0    0    0    0    0  997  997
 999  999  999  999  614  480    0    0    0    0    0    0    0  220 1000 1000
 999  999  999  999  480  480    0    0    0    0    0    0  991  220  847 1000
 999    0    0 1000  480  480    0    0    0  211  211  211  991  991  991  999
 106    0 1000 1000  857  857    0    0    0  211  211  211  991  991  220  220
```

(b)

```
   .    .    .    .    . 1000 1000 1000 1000    0    .    .    .    .    .    .
   .    .    .    .    .    0 1000    0    0 1000    .    .    .    .    .    .
   .    .    .    .    .    . 1000    0    0    0    .    .    .    .    .    0
   .    .    .    0    .    .    0    0 1000 1000 1000    0    .    .    .    .
1000 1000 1000    0    0    .    . 1000 1000    0    0    0 1000    0    .    .
1000 1000    0    0    0    .    .    .    0 1000    0    0 1000 1000    0
   0    0    0 1000    0 1000    .    .    .    0    0    0 1000    0 1000
   0 1000    0    0 1000    0    .    0    .    . 1000    0    0    0    0 1000
   . 1000    0 1000 1000 1000    0    0 1000    .    .    0    0 1000    0
   0 1000 1000 1000    0 1000    0    0    0    .    . 1000    0    0    .
   0    0    0    0    0    0    0    0 1000    . 1000    0    0 1000    .
   0    0    0    .    . 1000    0    0    0    0    0 1000    0 1000    .
   .    .    .    .    .    0 1000    0 1000    0 1000 1000    .    .    .
   .    .    .    .    . 1000 1000    0    0 1000    0    .    .    .    .
   . 1000 1000    .    .    .    0 1000    0    .    .    .    .    .
   . 1000    .    .    . 1000 1000    0    .    .    .    .    .
```

(c)

```
  6  80  80  80  80   0   0   0   0   0  15  15  80  80  80  15
  6  80  80  71  80   0   0   0   0   0  15  15  80  80  80   6
 80  80  71  71  80  80  80   0   0   0   0  15  80  80  80   0
 80  80  71   0  80  80   0   0   0   0   0   0  80  80  80  80
  0   0   0   0   0  68  68   0   0   0   0   0   0   0  80  80
  0   0   0   0   0  80  80  80  80   0   0   0   0   0   0   0
  0   0   0   0   0   0  80  80  80  80   0   0   0   0   0   0
  0   0   0   0   0   0  80   0  80  80   0   0   0   0   0   0
 80   0   0   0   0   0   0   0   0  80  80   0   0   0   0  80
  0   0   0   0   0   0   0   0   0  80  80   0   0   0  80  80
  0   0   0   0   0   0   0   0   0  80   0   0   0   0  80  80
  0   0   0  50  50   0   0   0   0   0   0   0   0   0  80  80
 80  80  80  80  50  39   0   0   0   0   0   0   0  15  80  80
 80  80  80  80  39  39   0   0   0   0   0   0  80  15  68  80
 80   0   0  80  39  39   0   0   0  15  15  15  80  80  80  80
  6   0  80  80  69  69   0   0   0  15  15  15  80  80  15  15
```

(d)

```
   .    .    .    .    .   80   80   80   80    0    .    .    .    .    .    .
   .    .    .    .    .    0   80    0    0   80    .    .    .    .    .    .
   .    .    .    0    .    .   80    0    0    0    .    .    .    .    0
   .    .    .    .    .    0    0   80   80   80    0    .    .    .    .
  80   80   80    0    0    .    .   80   80    0    0    0   80    0    .    .
  80   80    0    0    0    .    .    .    0   80    0    0   80   80    0
   0    0    0   80    0   80    .    .    .    .    0    0    0   80    0   80
   0   80    0    0   80    0    .    0    .    .   80    0    0    0    0   80
   .   80    0   80   80   80    0    0   80    .    .    0    0    0   80    .
   0   80   80   80    0   80    0    0    0    .    .   80    0    0    .
   0    0    0    0    0    0    0    0   80    .   80    0    0   80    .
   0    0    0    .    .   80    0    0    0    .    0    0   80    0   80    .
   .    .    .    .    .    0   80    0   80    0   80   80    .    .    .
   .    .    .    .    .   80   80    0    0   80    0    .    .    .    .
   .   80   80    .    .    .    0   80    0    .    .    .    .    .
   .   80    .    .    .   80   80    0    .    .    .    .    .
```

Some further definitions will allow us to design an algorithm to compute the forcing domain, an approximation of the stable core.

Definition

A boolean mapping $f:\{0, 1\}^n \to \{0, 1\}$ is *forcing* in its ith variable iff:

$$\exists x^*_i \in \{0, 1\}, \ \exists v^*_i \in \{0, 1\}: \ \forall y \in \{0, 1\}^n, \ y_i = x^*_i \Rightarrow f(y) = v^*_i$$

Variable i is called a *forcing variable* of mapping f; x^*_i the *forcing value* of variable i; v^*_i the *forced value* of mapping f for variable i.

This notion, first introduced by Kauffman [8], has proved very useful to characterise the stable elements in the iteration on F [2, 4]. Figure A.1.1(c) shows the forced value and forcing values for the boolean mappings in 2 variables.

The role played by forcing functions is crucial for stability. We have run simulations on boolean networks with a variable rate of forcing functions. They show that the stable core shrinks when this rate decreases, while the oscillating core increases (see Fig. A.1.4).

Figure A.1.3 (*opposite*) Stable core of the network of Fig. A.1.1 for parallel iterations (a) and sequential iteration with identical permutation (c). For each iteration mode, k different initial conditions have been tested ($k = 1000$ for the parallel iteration, $k = 80$ for the sequential mode) and thus k limit cycles computed. For each limit cycle and each automaton, the number N of times it was found oscillating is indicated, (a) and (c). For the elements always stable ($N = 0$), the values in which it was found stable have been added: points with total value 1000 (parallel) or 80 (sequential) were thus stable in the same value whatever the initial condition. This total value is shown in (b) and (d).

```
p = 0%
 0 16 16 15 16 14 16 16 16 16  0       16  .  .  .  .  .  .  .  .  .  .  16
 0 16 16 16  0 16 16 14 16  0  0  0    16  .  .  . 16  .  .  .  .  0 16 16
 0  0  0  0  0  0 16 16  0  0  0  0    16 16 16 16 16 16  .  .  0  0 16 16
 0  0  0  0  0  0 16 16 16  0  0  0     0  0  0 16 16  0  .  .  0 16 16
 0  0  0  0  0  0 16 16 16 16 16  0    16 16 16 16 16 16  .  .  .  .  .  0
 0  0  0  0  0  0  0  0  0 16 16 16    16  0  0 16 16 16 16  0 16  .  .  .
16  0  0  0  0  0  0  0  0 16  0 16     . 16 16  0 16 16  0  0 16  .  0  .
16 16  0  0 16 16 16 16  0  0  0 16     .  . 16 16  .  .  .  . 16  0  0  .
 0 11  0  0 16 16 16 16 16 10 10  0 14 16  .  0  0  .  .  .  .  .  0  .
11 11  0  0 16 16  0  0 16 10 10 14     . 16 16  . 16 16  .  .  .  .  .
11 11  0  0  0 16 14 16 16 10 10  0     .  0 16  0  .  .  .  .  .  .  0
11 16 16  0  0  0 16 16 16 16 16 11     .  .  . 16 16  0  .  .  .  .  .  .
T = 8.00        a = 0.47
```

```
p = 20%
 0 16 16 16 16  6  6 16 16 16 16 16    16  .  .  .  .  .  .  .  .  .  .  .
16 16 16 16 16  6  6 14  9 16 16 16     .  .  .  .  .  .  .  .  .  .  .  .
16 16  0  0  0 16  0 16 16 16 16 16     .  .  0  0 16  . 16  .  .  .  .  .
16 16  0  0  0 16 16 16 16 16 16 16     . 16  0  0  .  .  .  .  .  .  .  .
16 16  0  0  0  0 16 16 16 16 16 16     .  .  0  0 16 16  .  .  .  .  .  .
 0  0  0  0  0  0 16 16 16 16 16 16    16  0 16  0 16 16  .  .  .  .  .  .
16  0  0 16  0  0 16 16 16 16  0 16     . 16 16  . 16 16  .  .  .  . 16  .
16 15 15 16 16 16 16 16 16  0  0 16     .  .  .  .  .  .  .  .  .  0  0  .
 0  0 15 15 16 16 16 16 16 16 14 16    16 16  .  .  .  .  .  .  .  .  .  .
 0  0 15 16 16 16 16 16 16 16 16 16     0 16  .  .  .  .  .  .  .  .  .  .
 0  0  0  0  0 16 12 16 16 16 16  0     0  0  0 16  0  .  .  .  .  .  .  0
 0 16 16  0  0  0 15 16 16 16 14  0     0  .  . 16 16  0  .  .  .  .  . 16
T = 31.00           a = 0.69
```

```
p = 40%
15 13 13  0  0 16 16 16 16 16 16 15     .  .  . 16 16  .  .  .  .  .  .  .
14 13 16 16 16 16 16 16 16 14 15        .  .  .  .  .  .  .  .  .  .  .  .
15 15 16 16 16 16 16 16 16 16 15 15     .  .  .  .  .  .  .  .  .  .  .  .
14 15 16 16 16 16 16 16 16 15 15 15     .  .  .  .  .  .  .  .  .  .  .  .
14 13 16 16 16 16 16 16 15 15 15 15     .  .  .  .  .  .  .  .  .  .  .  .
15  0 16 16 16  0 15 15 15 15  2 15     . 16  .  .  . 16  .  .  .  .  .  .
15 16 16  0 16 16 16 16 16  1  1 16     .  .  . 16  .  .  .  .  .  .  .  .
 0  0  0  0  0 16 16 16 16 16 16 16    16 16  0 16  0  .  .  .  .  .  .  .
 0  0  0  0  0  0 16 16 16 16 16 16     0 16 16 16 16  0  .  .  .  .  .  .
 0  0  0  0  0  0 16 16 16 16 16 16     0  0  0 16 16  0  .  .  .  .  .  .
15 16 14  0  0 16 16 16 16 16 16 16     .  .  .  0  0  .  .  .  .  .  .  .
T = 160.00          a = 0.79
```

```
p = 50%
13 13 13 13 13 13 13 13 13 13 13 13     .  .  .  .  .  .  .  .  .  .  .  .
13 13 13 13 13 13 13 13 13 13 13 13     .  .  .  .  .  .  .  .  .  .  .  .
13 13 13 13 13 13 13 13 13 13 13        .  .  .  .  .  .  .  .  .  .  .  .
13 13 13 13 13 13 13 13 13 13 13 13     .  .  .  .  .  .  .  .  .  .  .  .
13 13 13 13 13 13 13 13 13 13 13 13     .  .  .  .  .  .  .  .  .  .  .  .
13 13 13 13 13 13 13 13 13 13 13 13     .  .  .  .  .  .  .  .  .  .  .  .
13  0 13 13 13 13 13 13 13 13 13 13     . 13  .  .  .  .  .  .  .  .  .  .
13 13 13 13 13 13 13 13 13 13 13 13     .  .  .  .  .  .  .  .  .  .  .  .
13 13 13 13 13 13 13 13 13 13 13 13     .  .  .  .  .  .  .  .  .  .  .  .
13 13 13 13 13 13 13 13 13 13 13        .  .  .  .  .  .  .  .  .  .  .  .
13 13  0  0 13 13 13 13 13 13 13 13     .  .  0 13  .  .  .  .  .  .  .  .
13 13 13 13 13 13 13 13 13 13 13 13     .  .  .  .  .  .  .  .  .  .  .  .
T = 1282.50         a = 0.98
```

Definition

Let $F: \{0, 1\}^n \to \{0, 1\}^n$ be a boolean network of connectivity $k = 2$ and let $\mathscr{C} = (X, \mathscr{W})$ be its connection graph. We define recursively two *labellings*

$$\ell: X \to \{0, 1\} \text{ of the nodes of } \mathscr{C}$$
$$v: \mathscr{W} \to \{0, 1, *\} \text{ of the arcs of } \mathscr{C}$$

as follows:

(i) If f_j is forcing in 2 variables i_1 and i_2 (i.e. $(i_1, j), (i_2, j) \in \mathscr{W}$), let v^*_j be its unique forced value and x^*_i be the forcing value of f_j for $i = i_1, i_2$ ($x^*_i = *$ if f_j is a constant mapping).
   Then we take: $\ell(j) = v^*_j$ and $v(\cdot, j) = x^*_i$ for $i = i_1, i_2$.
(ii) If f_j is not forcing, $\ell(j)$ and $v(i, j)$ are not defined.
(iii) If f_j is forcing in 1 variable i and if $\ell(i)$ has already been defined and $\ell(j)$ has not, then we take: $\ell(j) = f_j[\ell(i)]$ and $v(i, j) = \ell(i)$.

At the end of the process, ℓ is defined on a subset Y of X and v on a subset $\mathscr{W}'$ of $\mathscr{W}$. Then, arc (i, j) in $\mathscr{W}'$ is *forcing* iff: $i \in Y$ and $\ell(i) = v(i, j)$ or $v(i, j) = *$ (we then set $v(i, j) = \ell(i)$). Graph $\mathscr{F} = (Y, \mathscr{W}_f)$ is called the *forcing graph* of F iff $\mathscr{F}$ is the subgraph of $\mathscr{C}$ such that $(i, j) \in \mathscr{W}$ is in $\mathscr{W}_f$ iff (i, j) is forcing.

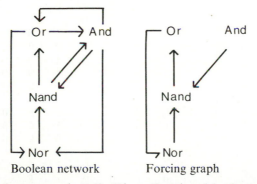

Boolean network   Forcing graph

Figure A.1.5 Boolean network $F: \{0, 1\}^4 \to \{0, 1\}^4$ and its forcing graph. Circuit $C = (\text{Or, Nor, Nand})$ is forcing and has been found, in simulations, stable in its forced value.

Figure A1.4 (*opposite*) Variable rate of forcing functions, for a boolean network with regular connections drawn on a torus; for a variable rate p of nonforcing functions the stable and oscillating cores, the stable value, the average period and individual activity (frequency of oscillation per automaton) for 16 different initial conditions are shown.

Any circuit of graph $\mathscr{F}$ is called a *forcing circuit* of graph $\mathscr{C}$. The notion of forcing circuit is related to the concept of *frustration* introduced in spin glasses [2]. Forcing circuits may also be characterised through the use of the discrete derivative introduced in Chapter 1 [2].

Let us now use this notion of forcing function to define the forcing domain of a boolean network.

Definition

$\mathscr{D} \in X$ is the *forcing domain* of network F and $v^*:\mathscr{D} \to \{0, 1\}$ is the *forcing value* of $\mathscr{D}$ iff:

(i) $\forall i \in Y$, i is on a forcing circuit $\Rightarrow i \in \mathscr{D}$ and $v^*(i) = \ell(i)$

(ii) $\forall i \in \mathscr{D}, \forall j \notin \mathscr{D}, (i, j) \in \mathscr{W}_f$ and $v^*(i) = v(i, j) \Rightarrow j \in \mathscr{D}$ et $v^*(j) = \ell(j)$

(iii) $\forall j \notin \mathscr{D}, \forall t(1, \ldots, k), i_t \in \mathscr{D}, (i_t, j) \in \mathscr{W} \Rightarrow j \in \mathscr{D}$ and $v^*(j) = f_j(v^*(i_1), \ldots, v^*(i_k))$

(iv) $\forall i \in \mathscr{D}, \forall j \notin \mathscr{D}$, if $(i, j) \in \mathscr{W}$ and $v^*(i)$ is a forcing value of f_j, then $j \in \mathscr{D}$ and $v^*(j)$ is the forced value of f_j associated with the forcing value $v^*(i)$.

```
  .   .   .   .   .   1   1   1   1   0   1   0   .   .   .   .   .
  .   .   .   .   .   0   1   0   0   1   0   .   .   .   .   .
  .   .   .   .   .   .   1   0   0   0   .   .   .   .   .
  .   .   .   .   .   .   0   1   1   1   0   .   .   .   .
  1   1   .   .   .   .   1   1   0   0   0   1   0   .
  1   1   0   0   0   .   .   .   0   1   0   0   1   1   0
  0   0   0   1   0   1   .   .   .   0   0   0   1   0   1
  0   1   0   0   1   0   .   .   1   0   0   0   0   1
  .   1   0   1   1   1   0   0   1   .   0   0   0   1   .
  .   .   1   1   0   1   0   0   0   .   1   0   0   .
  .   .   0   0   0   .   0   0   1   .   1   0   0   1   .
  .   0   0   .   .   .   .   0   0   0   1   0   1   .
  .   .   .   .   .   .   .   1   0   1   1   .
  .   .   .   .   .   .   .   0   1   0   .
  .   .   .   .   .   .   0   1   0   0   1   1   .
  .   .   .   .   .   .   1   1   0   0   1   0   .
```

Figure A.1.6 Forcing domain of the boolean network of Fig. A.1.1. Dots are points out of the domain, points in the domain are shown by their stable value 0 or 1.

The definition of the forcing domain explains how to propagate forced values along forcing arcs, starting from the forcing circuits.

Theorem 1

Let $F: \{0, 1\}^n \to \{0, 1\}^n$ be a boolean network of connectivity 2, v^* its stable value. Let C be a forcing circuit of F, then for any limit cycle of the parallel iteration on F:

either C is stable in value v^*/C

or C is oscillating, not in its stable value v^*/C and the following condition is satisfied:

$$\forall t(1, \ldots, T), \forall i, j \in \{1, \ldots, n\} \cap C, (i, j) \in \mathcal{W} \text{ and } s_i(t) \neq v^*_i \Rightarrow$$
$$\{\exists k(1, \ldots, n), k \neq i : (k, j) \in \mathcal{W} \Rightarrow s_k(t) \neq v(k, j)\} \quad \text{(A.1.1)}$$

where $s(t)$ is the state of the network at time $t(1, \ldots, T)$ of the limit cycle (of period T).

Theorem 2

Let $F: \{0, 1\}^n \to \{0, 1\}^n$ be a boolean network of connectivity 2, v^* its stable value. Let π be a permutation of $\{1, \ldots, n\}$ and C be a forcing circuit of F, then for any limit cycle of the sequential iteration F_π

either C is stable in value v^*/C

or C is oscillating with a period T and conditions (A.1.1) and (A.1.2) relating T and π are satisfied.

see [4] for condition (A.1.2).

Theorem 3

Let $F: \{0, 1\}^n \to \{0, 1\}^n$ be a boolean network of connectivity 2, v^* its stable value. Then for the parallel and sequential iterations on F, all forcing circuits are stable or they satisfy (A.1.1) (and (A.1.2) for the sequential iteration).

Consequence The two previous theorems imply that elements in the forcing domain will be stable, unless (A.1.1) (and (A.1.2)) are satisfied. The forcing domain is thus in general a good approximation of the stable core. Simulations have confirmed this result. However, it may happen that (A.1.1) (and (A.1.2)) are satisfied. In that case forcing circuits may be nonstable and the forcing domain different from the stable core [9].

This can be verified from Figs A.1.3 and A.1.6: the forcing domain (Fig. A.1.6) is very close to the stable core both for the parallel and sequential iterations (Fig. A.1.3).

A.1.4   Entropy in boolean networks

This section gives some results about the entropy, which allow the limit set of the dynamics to be characterised. These results are to be compared with Chapter 4 where an energy function allows the proof of similar, but more powerful results (because the energy is defined on *one* state, while the entropy is defined on the *whole* state space).

Definition

Let $F: \{0, 1\}^n \rightarrow \{0, 1\}^n$ be a boolean network and let p be a probability distribution on the state space $S = \{0, 1\}^n$.

The *entropy* of network F associated to p is the function E defined by:

$$E(p) = -\sum_{s \in S} p_s \log_2 p_s$$

where p_s is the probability of state $s \in \{0, 1\}^n$.

Theorem 4

Let p^0 be an initial probability distribution on S and F_d be an iteration mode on $F \cdot F_d$ defines a trajectory $p(t)$ with:

$$p(0) = p^0 \quad \text{and} \quad p_s(t) = \text{Proba}[x(t) = s]$$

Then: $\forall t \geq 1$,

$$E[p(t)] \leq E[p(t - 1)]$$

and $E[p(t)] = E[p(t - 1)]$ iff

$\forall s \in S$,

$$\text{either } F_d^{-1}(s) = \varnothing,$$
$$\text{or } \exists! \; \sigma \in F_d^{-1}(s) : p_\sigma(t - 1) \neq 0.$$

This theorem shows that entropy E is a Lyapunov function for the dynamics on p. It can also be used to study the dynamics of network F.

Theorem 5

Let $\mathscr{A}(t)$ be the set $\{s \in S : p_s(t) \neq 0\}$, that is the set of 'live' elements at time t.

Then $[\mathscr{A}(t)]_{t \geq 0}$ is a decreasing sequence (with respect to inclusion) of subsets of S, whose intersection $\mathscr{A} = \cap_{t \geq 0} \mathscr{A}(t)$ is nonempty: it is the *limit set* of the iteration, it contains all the states which are part of a limit cycle.

Furthermore, $E(t)$ is convergent and its limit E_∞ is such that: $\exists t^* \geq 1$:

$$E_\infty = -\sum_{s \in \mathscr{A}} p_s(t^*) \log_2 p_s(t^*)$$

This theorem thus allows the characterisation of those states which are part of limit cycles (see Fig. A.1.7).

Corollary 1

If S and σ are two states in $\mathscr{A}$ such that: $F_d(s) = \sigma$, then:

$$\forall t \geq t^*, p_\sigma(t + 1) = p_s(t)$$

Proof: obvious, because: $\forall s' \in F_d^{-1}(\sigma)$, $s' \neq s \Rightarrow s' \notin \mathcal{A} \Rightarrow p_s'(t + 1) = 0$.

$\square$

This corollary allows us to find out the structure of limit cycles and periods from the distribution $(p(t))_{t \geq t^*}$ (see Fig. A.1.7).

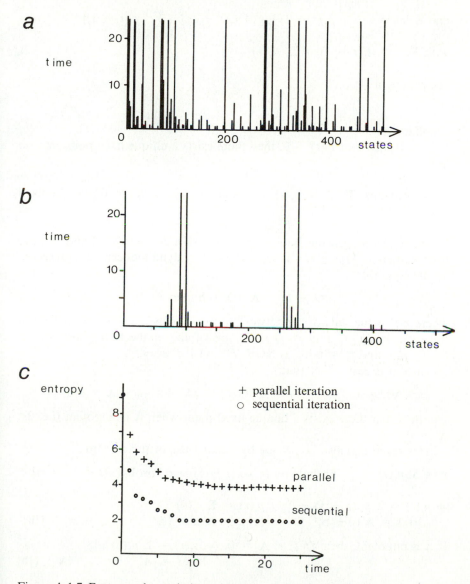

Figure A.1.7 Entropy: the variation with time of the probability of states for the parallel iteration (a) and the sequential iteration (b) and of the entropy (c).

Theorem 6
The transient length of F_d is bounded below:

$$\forall x^0 \in \{0, 1\}^n, \; t(x^0) \leq n/\varepsilon_F$$

with $\varepsilon_F = \min\{|E_s(t)|/t \geq 1, s \in S: F_d^{-1}(s) \cap \mathscr{A}(t - 1) \text{ has 2 elements at least}\}$

and $E_s(t) = \sum_{\sigma \in F_d^{-1}(s)} p_\sigma(t - 1) \log_2 [p_\sigma(t - 1) \Big/ \sum_{\tau \in F_d^{-1}(s)} p_\tau(t - 1)]$

with $E(t) - E(t - 1) = \Sigma_s E_s(t)$

See proofs in [4].

Corollary 2
If $\exists t > 0$ such that $E(t) = 0$, then there exists a unique fixed point and no limit cycle.

Proof: obvious: $\exists! \; s \in S: p_s(t) = 1$ and $\forall \sigma \neq s, \; p_\sigma(t) = 0$: s is the fixed point.   □

Example   Homogeneous boolean networks of mappings XOR and $\Leftrightarrow$ have been studied [6, 11] and it was shown that their dynamics can be represented by the equation:

$$x(t + 1) = \mathbf{A}x(t) + \mathbf{b} \quad (\text{in } \mathbb{Z}/2)$$

where $x(t)$ is the state of the network at time t, $\mathbf{A}$ is the incidence matrix of the network ($a_{ij} = 1$ iff (i, j) is an arc of the connection graph $\mathscr{C}$, i.e. $(i, j) \in \mathscr{W}$), and $b_i = 0$ if f_i is XOR, $b_i = 1$ if f_i is $\Leftrightarrow$.
    Then, it is easy to see that:

$$\forall t \geq 0, \quad x(t + 1) = \mathbf{A}^{t+1}x(0) + [\mathbf{A}^t + \ldots + \mathbf{A} + \mathbf{I}]\mathbf{b}$$

and thus that there exists a unique fixed point when $\mathbf{A}$ is nilpotent (i.e. $\exists r: \mathbf{A}^r = 0$).
    This result can also be proved by making use of the entropy:

$$\forall s \in S, p_s(t) = \sum_{\sigma \in S} \text{proba}[x(t) = s/x(t - 1) = \sigma] \, \text{proba}[x(t - 1) = \sigma]$$

$\Rightarrow p_s(t) = \Sigma \, p_\sigma(t - 1) \Rightarrow \qquad p_s(t) = \Sigma \, p_z(0)$
$ [\sigma: s = \mathbf{A} \, \sigma + \mathbf{b}\} \qquad \{z: s = \mathbf{A}^t z + [\mathbf{A}^{t-1} + \ldots + \mathbf{A} + \mathbf{I}]\mathbf{b}\}$

If $\mathbf{A}$ is nilpotent, then $\forall t \geq r: \mathbf{A}^t = 0, \Rightarrow p_s(t) = \Sigma \, p_z(0)$
$\phantom{\text{If } \mathbf{A} \text{ is nilpotent, then } \forall t \geq r: \mathbf{A}^t = 0, \Rightarrow p_s(t) =} \{z: s = [\mathbf{A}^{r-1} + \ldots + \mathbf{A} + \mathbf{I}]\mathbf{b}\}$

$$\Rightarrow \; p_s(t) = \begin{cases} 1 & \text{if } s = [\mathbf{A}^{r-1} + \ldots + \mathbf{A} + \mathbf{I}]\mathbf{b} \\ 0 & \text{otherwise} \end{cases}$$

$$\Rightarrow \forall t \geq r, \quad E(t) = 0$$

and thus there exists a unique fixed point: $s^* = [A^{r-1} + \ldots + A + I]b$ (Fig. A.1.8).

```
6 6 9 6 6 6 6 9     1 1 0 0 1 1 1 0        B D 0 0 0 0 0 D
6 9 9 9 9 9 6 6     0 0 1 1 1 0 0 0        E C E 0 0 0 0 0
6 6 9 9 6 9 6 6     1 0 0 1 1 0 1 1        0 D B D 0 0 0 0
9 9 6 6 9 9 9 9     0 0 1 1 0 1 0 1   A =  0 0 E C E 0 0 0
9 9 9 9 6 9 9 6     0 1 1 1 0 1 0 0        0 0 0 D B D 0 0
6 9 6 6 9 9 9 9     1 0 0 1 1 1 1 1        0 0 0 0 E C E 0
9 6 9 9 6 9 6 9     1 1 1 1 0 1 0 1        0 0 0 0 0 D B D
9 9 6 6 9 9 9 6     1 1 1 1 0 1 0 1        E 0 0 0 0 0 E C
```
with:

```
        01000001        00000000        00000000        10000000        00000000
        00000000        10100000        01000000        00000000        00000000
        01010000        00000000        00000000        00100000        00000000
B =     00000000   C =  00101000   D =  00010000   E =  00000000   0 =  00000000
        00010100        00000000        00000000        00001000        00000000
        00000000        00001010        00000100        00000000        00000000
        00000101        00000000        00000000        00000010        00000000
        00000000        10000010        00000001        00000000        00000000
```

Figure A.1.8 Boolean network F: $\{0, 1\}^n \rightarrow \{0, 1\}^n$, $n = 8 \times 8$, with regular connections (as in Fig. A.1.1) with mappings XOR (6) and $\Leftrightarrow$ (9) only: left. For 80 different initial conditions, set at random, the network always ended up in a fixed point (shown in the middle). The incidence matrix A of the network (right) is nilpotent: $A^8 = 0$.

A.1.5 Conclusion

We have presented here some results on the dynamic evolution of boolean networks. We have shown that networks with connectivity 2 have limit cycles whose geometrical structures are very similar. The notion of stable core has been shown to provide a good insight into the dynamics of the network. In fact the behaviour of these networks is very simple because the rate of forcing functions is high. Decreasing this rate (either by hand or by increasing the connectivity) leads to asymptotic behaviour far more unstable.

A.1.6 References

[1] E. Bienenstock, F. Fogelman Soulié and G. Weisbuch (eds), *Disordered Systems and Biological Organization*, NATO ASI Series in Computer and Systems Sciences, F20, Springer Verlag, Berlin, Heidelberg, New York (1986)

[2] F. Fogelman Soulié, Frustration and stability in random boolean networks, *Disc. Appl. Math.*, 9 (1984), 139–56

[3] F. Fogelman Soulié, Contributions à une théorie de calcul sur réseaux, Thesis, Grenoble (1985)

[4] F. Fogelman Soulié, Parallel and sequential computation on boolean networks, *Theoret. Comp. Sci.*, (1986)

[5] A. E. Gelfand and C. C. Walker, *Ensemble Modelling*, Dekker (1984)

[6] E. Goles Chacc, Comportement dynamique de réseaux d'automates, Thesis, Grenoble (1985)

[7] S. A. Kauffman, Behaviour of randomly constructed genetic nets, in *Towards a Theoretical Biology*, C. H. Waddington (ed.), Edinburgh University Press, 3, (1970), 18–37

[8] S. A. Kauffman, Boolean systems, adaptive automata, evolution, in [1], 339–60

[9] D. Pellegrin, Thesis, Grenoble (1986)

[10] F. Robert, *Discrete Iterations*, Springer Verlag, Berlin, Heidelberg, New York (1986)

[11] E. H. Snoussi, Structure et comportement itératif de certains modèles discrets, Thèse Docteur Ing., Grenoble (1980)

[12] R. Thomas (ed.) *Kinetic Logic, a Boolean Approach to the Analysis of Complex Regulatory Systems*. Lecture Notes in Biomathematics 29, Springer Verlag, Berlin, Heidelberg, New York (1979)

I am indebted to H. Atlan and G. Weisbuch for our fruitful collaboration.

From continuous to discrete systems: the theory of itineraries

2.1   Introduction

The great renaissance of qualitative dynamics (strange attractors, turbulence, chaos, transition from order to disorder, and so on) is in large measure due to the results obtained in the study of the iterative behaviour of one-dimensional systems (unimodal functions). The monographs of Collet and Eckmann [3] and of Gumowski and Mira [10] are excellent introductions to this domain of investigation.

A basic aspect in the study of the iterative behaviour of one-parameter families of unimodal functions is the combinatorial character of various properties. This character is related to the exchange of intervals. It could also be found in the study of continuous selfmaps of the circle [11], and more generally of a compact metric space.

The study of this combinatorial aspect can be done through coding techniques: it associates a discrete iteration to the continuous dynamical system. This discrete system can be considered as an automaton with memory or as an automata network. The iterative properties of this automaton lead to qualitative results on the dynamics of the original system (such as existence of cycles, of turbulent trajectories, of chaotic regions). Conversely, certain automata with memory can be transformed into continuous systems, Kitagawa [12], Cosnard and Goles [7].

This chapter presents the theory of itineraries in a general setting, then its applications in the case of unimodal functions and of continuous selfmaps of the circle. An example of an automaton with memory obtained by this theory is described in detail.

2.2   Simplified theory of itineraries

Let $J_1, \ldots, J_p$ be closed intervals of $\mathbb{R}$ (ordered increasingly) such that the intersection of two consecutive intervals contains at most one point. Let J

denote the union of the intervals J_j, $j = 1, \ldots, p$. We call a function f a *selfmap* of J if it statisfies the following conditions *:

* f is continuous on each J_j
* for all j there exists r and s such that f maps J_j onto $J_r \cup \ldots \cup J_s$.

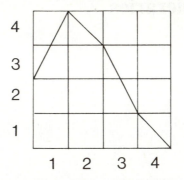

Figure 2.1 A function satisfying the selfmap conditions

Let j be the code corresponding to J_j. Denote by C the coding set: $C = \{1, \ldots, p\}$. The selfmap f maps an interval J_j to various intervals. We can summarise this behaviour by an oriented graph $G(f)$ constructed as follows.

C is the set of nodes of $G(f)$. There is an edge between node j and node r if J_r is included in $f(J_j)$. The graph $G(f)$ satisfies the following properties:

it possesses p nodes
each node is the start of at least one edge
all the edges starting from one node go to consecutive nodes

We shall denote by E the set of infinite sequences of elements of C and by $E(f)$ the subset of E constructed in the following way: $e = (e_j)$ belongs to $E(f)$ if there is an edge between e_j and e_{j+1} in $G(f)$. Hence $E(f)$ is the set of infinite paths of $G(f)$. The existence of at least one such path is deduced from the fact that f maps J into J.

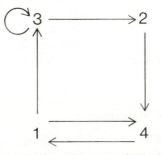

Figure 2.2 The graph associated to the function of Fig. 2.1.

Now let x belong to J. We denote by $\gamma(x)$ the element of C corresponding to the code of the interval containing x (since we have considered closed intervals, x can belong to two different intervals; hence we should associate to it a special code: for reasons of simplicity we shall not consider this problem). The itinerary $i_f(x)$ of x is the element of E defined as follows:

$$i_f(x) = (i_j) \text{ where } i_j = \gamma(f^j(x))$$

Hence the itinerary of x under f is the coding sequence of the intervals which contain the iterates of x. The set of itineraries under f of all the x in J will be called $I(f)$.

We define on $I(f)$ a distance δ as follows: the distance between two elements of $I(f)$ is the inverse of the rank of the first two different codes:

$$\delta(i, k) = 1/j \text{ if } i_1 = k_1, \ldots, i_{j-1} = k_{j-1} \text{ and } i_j \neq k_j.$$

The shift d maps i of $I(f)$ onto k of $I(f)$ where $i = (i_1, i_2, \ldots, i_j, \ldots)$ and $k = (i_2, i_3, \ldots, i_j, \ldots)$. It is easy to see that d is continuous with respect to the distance δ. The pair $(I(f), d)$ is called the discrete dynamical system associated to (J, f). The theory of itineraries studies the relationships between these two pairs. The great interest of this construction comes from the following result.

Lemma 1 For all n of $\mathbb{N}$, the itinerary of $f^n(x)$ is the itinerary of x shifted by n positions:

$$i_f(f^n(x)) = d^n(i_f(x)) \tag{2.1}$$

Let us now introduce some definitions.

Definitions

x of J (resp. $I(f)$) is called *n-periodic* if $f^n(x)$ (resp. $d^n(x)$) is equal to x while all the preceding iterates are different. x and its $n - 1$ iterates form a cycle of order n.

If n is equal to 1, x is called a *fixed point*.

x of J (resp. $I(f)$) is *asymptotically n-periodic* if $\{f^p(x)\}$ (resp. $\{d^p(x)\}$) can be decomposed into n subsequences which converge to the n points of a cycle.

An iterative sequence which is not asymptotically periodic is called *turbulent*.

Theorem 1

(i) If x of J is asymptotically *n-periodic*, then $i_f(x)$ is asymptotically *n-* or *n/2-periodic*.

If the itinerary $i_f(x)$ of x is *n-periodic* under d, then

(a) x is asymptotically *n-* or *2n-periodic* under f

(b) there exists y, n-periodic, such that $i_f(y) = i_f(x)$.

(ii) x is turbulent if and only if $i_f(x)$ is turbulent.

Hence the knowledge of iterative properties of d on $I(f)$ allows us to obtain precise information on the dynamics of f on J. Notice that the definitions of periodicity for $\{d^p(x)\}$ correspond exactly to the classical definitions. We have now to construct and study $I(f)$. It is a difficult problem. In the case we are dealing with, we can obtain the following result:

Theorem 2

The set of itineraries of f is equal to the set of infinite paths of $G(f)$:

$$I(f) = E(f)$$

Hence we can explicitly obtain the itineraries of f by constructing the graph $G(f)$ associated to f.

In the case of Fig. 2.1, whose graph is shown in Fig. 2.2, it is easy to see that $I(f)$ is the set of infinite sequences constructed using 1, 2, 3 and 4 along the following rules:

1 is followed by 3 or by 4

2 is followed by 4

3 is followed by 2 or by 3

4 is followed by 1

We deduce the existence of periodic itineraries of arbitrary length except 3 and an infinite number of turbulent sequences. Such a function is said to have a *chaotic behaviour*. Various properties can be obtained as for example the number of cycles of a given order. In this case, we have four cycles of order 9, namely $(13^624)^\infty$, $(13^42414)^\infty$, $(13^224(14)^2)^\infty$ and $(13^2241324)^\infty$. An example of a turbulent sequence is $32413^224133241 \ldots 3^j2413^{j+1}241 \ldots$

This example shows clearly the limitations of this theory. We can prove the existence of cycles and of turbulent sequences, we can construct itineraries with these properties, but this does not provide us with an efficient way to associate an x to a given itinerary.

2.3   Unimodal functions

A unimodal function f is a continuous selfmap of $J = [a, b]$ such that there exists c between a and b satisfying:

(i) $f(a) = \lambda$ arbitrary, $f(c) = b$ and $f(b) = a$

(ii) f is strictly increasing on $[a, c]$ and decreasing on $[c, b]$.

Let $J_0 = [a, c]$ and $J_1 = [c, b]$. With this partition, f does not satisfy the conditions * for selfmap on J_0. In some cases, a finer partition allows us to apply the preceding theory. However, such a partition does not exist in general. Indeed, f is one-to-one from J_1 into J, but is only injective from J_0 into J. The theory of itineraries is more complicated. A good introduction to the dynamics of unimodal maps can be found in [3].

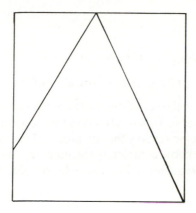

Figure 2.3 Graph of a unimodal function

Let x belong to J. As in the preceding paragraph, $\gamma(x)$ is equal to 0 (respectively 1) if x belongs to J_0 (respectively J_1). The itinerary of x is:

$$i_f(x) = (i_j) \text{ where } i_j = \gamma(f^j(x))$$

We denote by E the set of binary sequences $E = \{0, 1\}^{\mathbb{N}}$. The set $I(f)$ of itineraries is a subset of E. We introduce on E the distance δ and the shift d. We obtain in the same way the relation (2.1):

$$i_f(f^n(x)) = d^n(i_f(x)) \tag{2.1}$$

Theorem 1 can be proved without modification in this case. However, theorem 2 is no longer valid: we can only show that $I(f)$ is included in $E(f)$.

In order to more precisely study $I(f)$, we put on E the lexicographical ordering $\leq$, induced by $0 < 1$, and the associated topology. Let $e = (e_j)$ belong to E. Let e' be the complement of e: $e'_j = 1 + e_j$, where the addition is taken in $\mathbb{Z}/2\mathbb{Z}$. We define on E the function t: $te = (te_j)$ with $te_j = e_1 + \ldots + e_j$. The following property is of basic importance.

Lemma 2
Let x and y belong to J. If $x < y$ then $ti_f(x) \leq ti_f(y)$ (2.2)

The use of t comes from the fact that f is decreasing on J_1. We shall see in the following paragraph that, for an increasing function, t is not useful.

Relation (2.2) has important consequences for the characterisation of

$I(f)$. Since f maps J into itself, we obtain that, for all x in J, $f''(x) \leq b$. Using (2.1) and (2.2), we deduce that:

$$ti_f(f''(x)) = td''i_f(x) \leq ti_f(b) \tag{2.3}$$

Indeed, the following theorem states that this necessary condition is also sufficient.

Theorem 3
An element i of E belongs to $I(f)$ if and only if:

$$td''i \leq ti_f(b) \quad \text{for all } n \text{ of } \mathbb{N}.$$

Theorem 3 implies that the set of itineraries of f is characterised by a unique element, the itinerary of b. This result has in this case the same importance as Theorem 2. However, we remark that its practical use is difficult since it is in general impossible to know completely the itinerary of b under f. It can have some theoretical consequences. For example, we can deduce the following theorems.

Theorem 4
If there exists z in J such that $f^3(z) \leq z < f(z) < f^2(z)$ then f has a cycle of order k, for all integers k, and an infinite number of turbulent sequences.

This theorem was first proved by Li and Yorke [13] in a more complete form and in a more general setting (we have only to assume that f is continuous). It has a great historical importance in the development of research in the dynamics of one-dimensional systems.

Theorem 5
Consider the following total ordering α:

$$1 \, \alpha \, 2 \, \alpha \, 4 \, \alpha \, \ldots \, \alpha \, 2^i \, \alpha \, 2^{i+1} \, \alpha \, \ldots$$

$$\ldots \qquad \ldots \qquad \ldots \qquad \ldots$$

$$\ldots \, \alpha \, 2^r(2p + 1) \, \alpha \, 2^r(2p - 1) \, \alpha \, \ldots \, \alpha \, 2^r.5 \, \alpha \, 2^r.3$$

$$\ldots \, \alpha \, 2^{r-1}(2p + 1) \, \alpha \, 2^{r-1}(2p - 1) \, \alpha \, \ldots \, \alpha \, 2^{r-1}.5 \, \alpha \, 2^{r-1}.3$$

$$\ldots \qquad \ldots \qquad \ldots \qquad \ldots$$

$$\ldots \, \alpha \, 2(2p + 1) \, \alpha \, 2(2p - 1) \, \alpha \, \ldots \, \alpha \, 10 \, \alpha \, 6$$

$$\ldots \, \alpha \, (2p + 1) \, \alpha \, (2p - 1) \, \alpha \, \ldots \, \alpha \, 5 \, \alpha \, 3$$

(i) If f admits a cycle of order k, then, for every u such that $u \, \alpha \, k$, f has a cycle of order u.

(ii) If f admits a cycle of order k, f does not always admit a cycle of order v, for every v such that $k \, \alpha \, v$.

Theorem 5 was originally proved by Sarkovskii [16] in a more general setting (for continuous functions). It contains as a particular case a simplified version of Theorem 4: the existence of a cycle of order 3 implies the existence of cycles of all orders. A detailed proof of this result can be found in Cosnard [5].

Another consequence of Theorem 3 is the following result on the existence of chaotic functions. This result is proved in Cosnard [4].

Theorem 6

(i) If f has no cycles of order 2^i, for all i of $\mathbb{N}$, then f has no turbulent sequence and every element of J is asymptotically periodic.

(ii) If f has a cycle of order not a power of 2, then f has an infinite number of cycles and turbulent sequences: f has a chaotic behaviour.

Theorem 3 shows the importance of the itinerary of b. In fact, it does not completely characterise f. But two functions with the same itinerary for b have common properties: they are macroscopically conjugate [6]. The following result shows that the fact that an element of E is the itinerary of b for a unimodal function is a purely combinatorial property.

Theorem 7

Let i belong to E. There exists f unimodal such that $i = i_f(b)$ if and only if one of the two equivalent conditions is satisfied:

(i) $td^n(i) \leq ti$ for all n of $\mathbb{N}$
(ii) $(ti)' \leq d^n(ti) \leq ti$ for all n of $\mathbb{N}$.

where $(ti)'$ denotes the complement of ti.

Note that the second condition works only on ti. If we set $k = ti$, we obtain: $k' \leq d^n k \leq k$. The set K of the sequences k characterises the itineraries $i_f(b)$ for every unimodal function. The study of K has been partially done by Allouche and Cosnard [1] who have shown that this set has a fractal structure.

2.4   The functions of the circle

Let $J_0 = [a, c]$, $J_1 = [c, b]$ and $J = [a, b]$. Let f be a selfmap of J such that:

(i) $f(a)$ and $f(b)$ are arbitrary, $f(c^-) = b$ and $f(c^+) = a$
(ii) f is continuous strictly increasing on J_0 and J_1

Note that f is a continuous increasing selfmap of the circle. To show this we identify a and b with a point of the circle of length $b - a$. This point, called 0, is the origin. F is defined by: $F(x) = f(a + x) - a$. Clearly F has no discontinuity in c.

In the following, we shall distinguish two cases: f onto ($f(a) > f(b)$) or into ($f(a) < f(b)$). The limit case ($f(a) = f(b)$) corresponds exactly to the

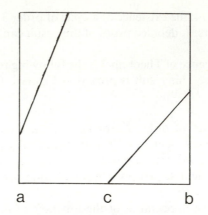

a   c   b

Figure 2.4 Example of a function of the circle

iteration of a diffeomorphism of the circle, which has been studied classically by Denjoy [8].

For x in J, we define the rotation number $\rho(x)$ as:

$$\rho(x) = \lim_{n \to \infty} (1/n) \sum_{j=1}^{n} \gamma(f^j(x))$$

where γ is defined in the preceding section. The following result can be found in Keener [11].

Theorem 8
If $f(a) > f(b)$, then

(i) $\rho(x)$ exists and is independent of x
(ii) If $\rho(x)$ is rational, x is asymptotically periodic
(iii) If $\rho(x)$ is irrational, x generates a turbulent sequence whose limit set is a Cantor set
(iv) If f belongs to a family of functions depending regularly on a parameter λ, then $\rho(x, \lambda)$ is a Cantor function of λ (called the devil's staircase), which takes rational values on nonempty intervals and irrational values on a Cantor set of measure zero

The case $f(a) < f(b)$ is more complicated. In general, the rotation number $\rho(x)$ varies in an interval and the iterative behaviour of f is chaotic.

Note the relationship between the rotation number of x and its itinerary: $\rho(x)$ measures the proportion of 1 in $i_f(x)$. The rotation number $\rho(x)$ is rational (resp. irrational) if and only if $i_f(x)$ is asymptotically periodic (resp. turbulent).

2.5 From itineraries to automata

The following automaton with memory has been introduced as a neuron model [2, 12]:

$$y_{n+1} = 1\left(\sum_{i=0}^{p} \alpha_i y_{n-i} - \beta\right) \tag{2.4}$$

where y_n belongs to $\{0, 1\}$, α_i and β are real parameters, p (the size of the memory) is either equal to n (bounded memory), or to a given integer (unbounded memory) and the function 1 is defined as:

$$1(u) = \begin{cases} 1 & \text{if } u \geq 0 \\ 0 & \text{otherwise} \end{cases}$$

Some results on the dynamical behaviour of this automaton have been obtained by Nagumo and Sato [15] and Cosnard and Goles [7]. In this paragraph, we shall be concerned with the case $\alpha_i = -\alpha^i$ where $0 < \alpha < 1$ and β is arbitrary.

The case of a bounded memory (p is fixed independently of n) will be studied in Chapter 5. Hence we restrict ourselves to the case of an unbounded memory ($p = n$):

$$y_{n+1} = 1\left(\beta - \sum_{i=0}^{n} \alpha^i y_{n-i}\right) \tag{2.5}$$

We shall prove that this automaton generates the itinerary of a particular point under the iteration of a function of the circle. We make the following change of variable:

$$x_n = \beta - \sum_{i=0}^{n} \alpha^i y_{n-i} \tag{2.6}$$

(2.5) is equivalent to: $y_{n+1} = 1(x_n)$ and we have:

$$x_{n+1} = \beta - \sum_{i=0}^{n-1} \alpha^i y_{n+1-i}$$

$$= \alpha x_n + (1 - \alpha)\beta - y_{n+1}$$

The change of variable leads to the following recurrence relation:

$$x_{n+1} = \alpha x_n + (1 - \alpha)\beta - 1(x_n) \tag{2.7}$$

It is not difficult to see that (2.7) is the equation of a linear injection of the circle whose iterative behaviour has been studied in Section 2.4.

Conversely, if we set $a = (1 - \alpha)\beta - 1$ and $b = (1 - \alpha)\beta$, the function $f(x) = \alpha x + (1 - \alpha)\beta - 1(x)$ maps $J = [a, b]$ into itself. We define $c = 0$, $J_0 = [a, c]$ and $J_1 = [c, b]$. We can show easily that the itineraries of β and of $\beta - 1$ are given by (2.5):

(i) $i_f(b) = (y_j)$ if $y_1 = 1$
(ii) $i_f(b - 1) = (y_j)$ if $y_1 = 0$

If we apply the preceding results, we can describe the dynamical behaviour of (2.7) and hence of (2.5). See Figure 2.5.

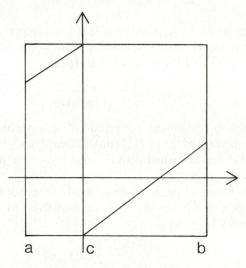

$$a \qquad\qquad c \qquad\qquad\qquad\qquad b$$

Figure 2.5 Graph of $f(x) = \alpha x + (1 - \alpha)\beta - 1(x)$

Theorem 9
For all β, (2.5) and (2.7) have a unique attractor which is either a cycle, or a Cantor set.
For all T, there exists β such that (2.5) and (2.7) have a T-cycle.

Let us define the rotation number associated to (2.5) and (2.7). It is equal to:

$$\rho_\alpha(\beta) = \lim_{n \to \infty} (\Sigma y_i / n)$$

Clearly, if $\rho_\alpha(\beta)$ is rational, then (2.4) has an attractive cycle and, otherwise, the attractor of (2.4) is a Cantor set.

Theorem 10
$\rho_\alpha(\beta)$ is a continuous increasing function whose derivative is almost everywhere zero (devil's staircase).

2.6 Conclusion

To study the iterative behaviour of a system of ordinary differential equations, a method, introduced by Poincaré, consists in intersecting the trajectories of the system with a hyperplane and in considering the successive intersections: we obtain a selfmap of the hyperplane. Hence this method transforms a continuous n-dimensional iteration with continuous time into a continuous $(n - 1)$-dimensional iteration with discrete time. From a practical point of view it could have some interest: a lower dimension allows an easier representation and the discrete time more precise and faster computations. However, the cases for which we can derive the associated Poincaré map explicitly are very rare. From a theoretical point of view, it is of basic importance: the tools used in studying such maps are conceptually simpler than those used for differential equations.

The situation is very similar in the case of the theory of itineraries. In fact this theory allows the transformation of the study of a continuous function with discrete time into the study of a discrete iteration. However, in practice, it has only limited interest since the theory does not lead explicitly to the correspondance between an itinerary and a point. From a theoretical point of view, it is rather different. The theory of itineraries allows us to know if a function has a chaotic behaviour or not, and to study the bifurcation structure of one-parameter families of maps.

Conversely, automata with memory can have a behaviour equivalent to continuous iterations. A complete study of these automata and of relationships between discrete systems with memory and continuous systems remains to be done.

2.7 References

[1] J. P. Allouche and M. Cosnard, Itérations de fonctions unimodales et suites engendrées par automates, *C. R. Acad. Sci.*, 296 (1983), 159–62
[2] E. R. Caianiello and A. De Luca, Decision equation for binary systems. Application to neuronal behavior, *Kybernetik*, 3 (1966), 33–40
[3] P. Collet and J. P. Eckmann, *Iterated Maps of the Interval as Dynamical Systems*, Birkhäuser, Basel (1980)
[4] M. Cosnard, On the behavior of successive approximations, *SIAM J. Numer. Anal.*, 16 (1979), 300–10
[5] M. Cosnard, Contributions à l'étude du comportement itératif des transformations unidimensionnelles, Thesis, Grenoble (1983)
[6] M. Cosnard, Etude de la classification topologique des fonctions unimodales, *Annales de l'Institut Fourier* 35, 3 (1985), 59–77
[7] M. Cosnard and E. Goles, Dynamique d'un automate à mémoire modélisant le fonctionnement d'un neurone, *C. R. Acad. Sci.*, 299, 10 (1984), 459–61
[8] A. Denjoy, Sur les courbes définies par les équations différentielles à la surface du tore, *J. Math. Pures Appl.*, 11 (1932), 339–52
[9] C. Gillot, Structure des itinéraires symboliques et de l'entropie topologique des transformations unimodales, Thesis, Toulouse (1984)
[10] Gumowski and C. Mira, *Dynamique chaotique*, CEPADUES Editions, Toulouse (1981), English translation: Manchester University Press (1986)
[11] J. P. Keener, Chaotic behavior in piecewise continuous difference equations, *Trans. Amer. Math. Soc.*, 261, 2 (1980), 589–604

[12] T. Kitagawa, Dynamical systems and operators associated with a single neuronic equation, *Math. Biosc.*, 18 (1973)

[13] T. Y. Li and J. A. Yorke, Period three implies chaos, *Amer. Math. Monthly*, 82 (1975), 985–96

[14] H. Nagami *et al.*, Characterization of dynamical behaviour associated with a single neural equation, *Math. Biosc.*, 18 (1973)

[15] J. Nagumo and S. Sato, On a response characteristic of a mathematical neuron model, *Kybernetik*, 3 (1972), 155–64

[16] A. N. Sarkovskii, Coexistence of the cycles of a continuous mapping of the line into itself, *Ukrain. Mat. Z.*, 16 (1964), 61–71

3 *Jacques Demongeot*

Random automata networks

3.1 Introduction

The theory of random automata networks is recent: its study began with
works by Rabin [24], Arbib [2] and Paz [22], following papers on stochastic
sequential machines like [5] and [25]. More recently, Doberkat [13] and
Milgram [21] have obtained more precise results concerning the dynamical
behaviour of these automata, especially asymptotic properties. The relation
between random automata networks and random fields [23, 26], is currently
being studied [15, 10, 11, 12]. Many applications have been proposed in
image processing [6, 15, 16], neural modelling [1, 4, 19, 20], epidemiology
[27, 8, 9, 10] and physics [28, 29]. Statistical tools also exist [3, 7, 18] allowing
estimation of the parameters for random automata networks. In Section 3.2
we give some definitions about random automata networks and random
fields; in Section 3.3 we show three examples of random automata networks:
networks with random iteration [14], binary Markovian and renewal
networks. In Section 3.4 we give some dynamical properties for these
examples, by using results on stability of attractors in these random
dynamical systems; these results are based on the notion of energy function
[14] or entropy [16]. Finally, we show some simulations in the renewal case.

3.2 Definitions

3.2.1 *Definition 1*

A *random automaton* A on a triplet (Q, Y, Z) of finite sets is defined by giving
two functions F and G, where Q is the state space, Y is the set of inputs, and Z
is the set of outputs:

$F: Y \times Q^2 \rightarrow [0, 1]$ denotes the probability that we go from q to q', when
the input is y; F verifies:

$$\forall y \in Y, \forall q \in Q, \sum_{q' \in Q} F(y, q, q') = 1$$

$G: Q \times Z \to [0, 1]$ is the probability that z is the output corresponding to the state q; G verifies:

$$\forall q \in Q, \sum_{z \in Z} G(q, z) = 1$$

3.2.2 Definition 2

A *random automata network* S on the set of sites X is defined by giving two functions R and C:

  $R: X \to \mathbf{A}$ allows the definition, in each site x of X, of an automaton $R(x)$, where $\mathbf{A}$ denotes the set of all automata on (Q, Y, Z)
  $C: X \times Z^X \to Y$ denotes the connection map of the network S, that is, $C(x, f)$ is the input received in the site x, when each site $u \in X$ gives the output $f(u)$.

The automaton $R(x)$ can be the same at each iteration and site x (parallel network) or can depend on the time of iteration and on the site x (sequential network); this dependence can be random.

3.2.3 Definition 3

A *random field* $\mathscr{C}$ on X is a set of random variables $\{C_x\}_{x \in X}$, such that there exists a probability space $(\Omega, \mathscr{A}, P)$ with:

$$\forall x \in X, \mathscr{C}: (\Omega, \mathscr{A}, P) \to (Q^X, B(Q^X), P_{\mathscr{C}}),$$

where $B(Q^X)$ denotes the Borelian σ-algebra on Q^X generated by finite cylinders and $P_{\mathscr{C}}$ denotes the canonical measure of $\mathscr{C}$.

3.3 Examples

3.3.1 Networks with random iteration

In [14], F. Fogelman Soulié studies *networks with random iteration*, that is, deterministic networks for which F and G are Dirac distributions, but R depends on the time of the iteration in the following way: by using a random process $\{I_n\}_{n \in \mathbb{N}}$, X-valued, we define R at the iteration n, such that:

  $F(I_n)$ is equal to F
  $\forall x \neq I_n, \forall q \in \{0, 1\}, \forall y \in Y, F(x)(y, q, q) = 1$

Such a network corresponds to a deterministic sequential network with a random choice of the successive sites x_n in which the automaton $R(x_n)$ changes the state $q(x_n)$, the other sites remaining in the same state.

In the following examples, the randomness will appear in F and G, but the choice of successive sites x_n will be deterministic. By melting the two possible randomisations, we obtain the more general case of random automata networks.

3.3.2   *Binary Markovian and renewal networks*

A *binary network* is defined by:

(i) $\forall x \in X$, $Q(x) = Z(x) = \{0, 1\}$

(ii) $X \subset \mathbb{Z}^d$, $|X| < +\infty$

(iii) $Y(x)$, the set of inputs for the automaton $R(x)$, is equal to $\{0, 1\}^{N(x)}$, where $N(x)$ represents a certain neighbourhood of x.

(iv) $\forall x \in X$, $G(x) = G$, where

$$G(q, z) = 1 \quad \text{if } z = q$$
$$\qquad\quad = 0 \quad \text{if } z \neq q$$

(v) If $Q = \{0, 1\}$, we can identify Q^X and $P(X)$, the set of all subsets of X. Then $B(P(X))$ is the set of all subsets of $P(X)$ and any probability $P_\mathscr{C}$ on $P(X)$ can be defined as a *Gibbs measure* by a potential $U: P(X) \rightarrow \mathbb{R}$, with $U(\varnothing) = 0$, in the following way:

$$\forall A \subset X, \; P_\mathscr{C}(\{A\}) = e^{U(A)}/Z$$

where $Z = \sum_{D \subset X} e^{U(D)}$ is a normalisation constant.

We can prove that U is defined from an associated interaction $V: P(X) \rightarrow \mathbb{R}$ verifying:

$$\forall A \subset X, \; U(A) = \sum_{D \subset A} V(D).$$

Then $F(x)$ is said to be defined from the potential U if the following formula holds (it is obtained after simplification of the normalisation constants):

$$\forall q \in \{0, 1\}, \; \forall D \subset N(x), \; F(x)\,(D, q, 1) = \frac{e^{U(D \cup \{x\})}}{e^{U(D \cup \{x\})} + e^{U(D)}}$$

This $F(x)$ $(D, q, 1)$ is simply the conditional probability that we observe the configuration $D \cup \{x\}$, given that input is D in $N(x)$, the basic probability being $P_\mathscr{C}$.

(vi) $C(x, f) = \{f(u)\}_{u \in N(x)}$

(vii) R is sequential and the dynamics of iteration is defined from an ordering $\{x_1, \ldots, x_{|X|}\}$ on X at the iteration $k|X| + i$ (where $k \in \mathbb{N}$) by:

(a) $F(x_i)$ is defined from the potential U

(b) $\forall x \neq x_i, q \in \{0, 1\}, \forall D \subset N(x), F(x)(D, q, q) = 1$

The binary network is called *Markovian with range B*, if we have:
(i) $N(x)$ is defined by:
   $N(x) = (x + B)\setminus\{x\}$, where B is a subset of $\mathbb{Z}^d$ containing the origin O.
(ii) the probability $P_\mathscr{C}$ verifies, if $[A] = \{E; A \subset E \subset X\}$:
   $\forall A, D \subset X, P_\mathscr{C}([A]|[D]) = P_\mathscr{C}([A]|[D \cap N(A)])$,
   where $N(A) = \bigcup_{x \in A} N(x)$

This property is ensured by an interaction V verifying ([7, 23]):

$$V(D) = 0, \text{ if } \forall x \in X, D\Delta N(x) \neq \varnothing$$

The binary network is called *renewal*, if the probability $P_\mathscr{C}$ verifies:

$$\forall A, D \subset X, P_\mathscr{C}([A]|[D]) = P_\mathscr{C}([A]|[\Pi_A(D)]),$$

where $\Pi_A(D) = \pi_A(D) \cap N(\pi_A(D)) \cup \{x \in D; d(x, A) = d(D, A)\}$
and $\pi_A(D) = \{x \in D; d(x, A) \leq d(D, A) + \sup_{z \in B} d(0, z) - 1\}$.

This property is ensured by an interaction V verifying [11]:

$$\exists j \in \mathbb{R}; V(D) = (-1)^{|D|} j, \text{ if } \forall x \in X, D\Delta N(x) \neq \varnothing$$

The Markovian and the renewal properties above are the spatial homologues of the one-dimensional notions of bidirectional Markovian and renewal character. They express the fact that local rules of the automata depend only on a certain neighbourhood of the site to be changed, in the Markovian case, or only on its nearest sites having the state 1 in the renewal case; it is easy to conceive that these two properties can be relevant in short interaction processes or in contagion processes.

3.4 Dynamical properties

3.4.1 *Networks with random iteration*

In [14] there is defined the notion of *stable* and *oscillating core* of the network: for any realisation of the random process $\{I_n(\omega)\}$, the dynamics of iteration leads to an asymptotic state where, for the subset $SC(\omega)$ of X, all sites keep the same state and, for the subset $X\setminus SC(\omega)$, all sites have an oscillating state. Then the stable core SC is defined by:

$$SC = \bigcap_{\omega \in \Omega} SC(\omega)$$

and the oscillating core by

$$OC = \bigcap_{\omega \in \Omega} (X\setminus SC(\omega)).$$

SC and OC are robust with respect to perturbations [14], hence it is interesting to study these two remarkable sets, either theoretically or through simulations.

For any configuration of states $q^* = \{q(x)\}_{x \in X}$, we can define an *energy-like function* $E(q^*)$, by

$$E(q^*) = U(\{x \in X; q(x) = 1\})$$

It is interesting to look at the possible monotonicity of $E(q)$, during an iteration dynamics $\{I_n(\omega)\}$. Let us suppose now that the function $F(x)$ is a *threshold function*:

$$\forall x \in X, \, \exists b(x) \in \mathbb{R}; \, \forall D \subset N(x), \, \forall q \in Q,$$
$$F(x)(D, q, 1) = 1 \quad \text{if } 2 \cdot \sum_{y \in D} V(\{x, y\}) \leq -(b(x) + V(\{x\}))$$

$$= 0 \quad \text{elsewhere.}$$

When $N(x)$ corresponds to the nearest neighbourhood of x, it is proved in [14] that the function $E(q^*)$ decreases during random iteration, when it is equal to:

$$E(q^*) = \sum_{x \in X} q(x) \sum_{y \in N(x)} V(\{x, y\}) q(y) + \sum_{x \in X} (b(x) + V(\{x\})) q(x)$$

The condition on E for decrease during random iteration is:

$$\forall x \in X, \, V(\{x\}) \leq 0$$

Let us suppose that we have more:

$$\forall x, y \in X, \, \forall \, q^* \in Q^X, \, \left(\frac{\partial E(q^*)}{\partial q(y)} - 2 \, V(\{x, y\})(2q(x) - 1) \right) (2q(x) - 1) \leq 0$$

Then, if the two conditions above are fulfilled, the dynamics of iteration has only fixed points; such a fixed point q^* is a local minimum of energy in its first neighbourhood, that is, is in the set of configurations in Q^X which differ from q^* only in a site x of X.

Moreover, this first neighbourhood is stable under the random iteration. When the process $\{I_n\}$ is visiting all sites of X, then q^* is attractive in its first neighbourhood [14], and $\| \overrightarrow{\text{grad}} \, E(q^*) \|$ characterises the degree of stability of q^*. This stability criterion is similar to that obtained from entropy in the case of binary Markovian and renewal networks below. Let us note finally that q^* belongs to the stable core SC.

3.4.2 *Binary Markovian and renewal networks*

We show in [10] and [11] that there exists a unique invariant measure for the Markov chain, $P(X)$-valued, whose successive states are the configurations observed at times of iteration $0, |X|, 2|X|, \ldots, k|X|, \ldots$; the transition matrix of this Markov chain, denoted by M, is defined as follows

$$M = \prod_{i=1}^{|X|} M_i,$$

where M_i corresponds to the transition between the iteration $k|X| + i - 1$ and the iteration $k|X| + i$; if $(M_i)_{D,E}$ denotes the general coefficient of the matrix M_i (where $D, E \subset X$), that is the probability that we go from the state D at iteration $i - 1$ to the state E at iteration i, then we have:

$$(M_i)_{D,E} = 0 \quad \text{if } |D\Delta E| > 1$$
$$= F(x_i)(D \cap N(x_i), \mathbf{1}_D(x_i), \mathbf{1}_E(x_i)) \text{ elsewhere.}$$

It is easy to verify that, if F is defined from the potential U associated to the measure $P_{\mathscr{C}}$, then $P_{\mathscr{C}}$ is invariant for M_i, hence for M. The general coefficient of M, denoted by $M_{D,E}$ verifies:

$$M_{D,E} = \prod_{i=1}^{|X|} (M_i)_{D_i, D_{i+1}}$$

where $\quad D_1 = D$
$$\vdots$$
$$D_i = (D\backslash((D\backslash E) \cap A_i)) \cup ((E\backslash D) \cap A_i), \text{ with } A_i = \{x_1, \ldots, x_{i-1}\}$$
$$\vdots$$
$$D_{|X|} = E$$

Because of the strict positivity of the coefficients of M, this matrix is irreducible and primitive, hence $P_{\mathscr{C}}$ is the unique invariant measure of the Markov chain associated to M. We say that the network realises asymptotically the random field $\mathscr{C}$: the network allows the simulation of $P_{\mathscr{C}}$ and for this reason is called a *Gibbs sampler*. In [15] and [17], there is studied a Gibbs sampler in which the iteration is random: such a network corresponds to an intermediary case between the networks defined in Sections 3.3.1 and 3.3.2 and represents the more general case of random automata networks.

We can define now the Kolmogorov–Sinaï *entropy* of the Markov chain by:

$$H = -\sum_{D \subset X} P_{\mathscr{C}}(\{D\}) \sum_{E \subset X} M_{D,E} \log (M_{D,E})$$

Let us denote by δ the Kullback distance; then we have, for any initial measure ν on X:

$$\delta(\nu M^k, P_{\mathscr{C}}) \leq e^{-kH}, \text{ for } k \text{ sufficiently large [16]}.$$

The entropy H gives a criterion for the convergence of νM^k to the invariant measure $P_{\mathscr{C}}$, hence H characterises the stability of the attractive measure $P_{\mathscr{C}}$. We can note also that $P_{\mathscr{C}}$, because of its Boltzmann character, maximises its own entropy among Gibbs measures having the same mean energy.

Remark The extension to an infinite network (where X is an infinite subset of $\mathbb{Z}^d$) is possible through a procedure similar to the thermodynamic limit [11]. The problem of phase transition can be solved in certain simple cases.

3.5 Simulations

We shall now consider a simulation of a renewal network. The indexation $x_1, \ldots, x_{|X|}$ of X corresponds to a spiral enumeration; the parameters given for each iteration correspond to:

(i) the iteration number
(ii) the number of sites in state 1 at this step
(iii) $N00_i$ (resp. $N01_i$), the number of sites whose state changes from 0 to 0 (resp. from 0 to 1) between steps $i_{|X|} - 1$ and $i_{|X|}$
(iv) $N11_i$ (resp. $N10_i$), the number of sites whose state changes from 1 to 1 (resp. from 1 to 0) between steps $i_{|X|} - 1$ and $i_{|X|}$

A relative convergence criterion on these parameters allows us to decide the end of the transient; after this transient phase, the network gives configurations of the Gibbs measure corresponding to the potential U associated to the function F of the automata. The network has the same behaviour as a Gibbs sampler and this property is used to generate random configurations with a given Gibbs distribution; it is particularly interesting to simulate *a priori* Bayesian Gibbs measures in image processing [15, 17] and in study of epidemiological processes; in this last case, the Harris contact process [10] can be simulated by the network, because both have the same invariant measure.

The simulations presented in Figures 3.1 to 3.5 where obtained on a VAX 11/780.

```
1 1 1 1 1 1 1 1 1 1 1
1 1 1 1 1 1 1 1 1 1 1
1 1                 1 1
1 1                 1 1
1 1                 1 1
1 1                 1 1
1 1                 1 1
1 1                 1 1
1 1                 1 1
1 1 1 1 1 1 1 1 1 1 1
1 1 1 1 1 1 1 1 1 1 1
```

Figure 3.1 Simulation of a renewal network: iteration number = 1; number of 1 = 72

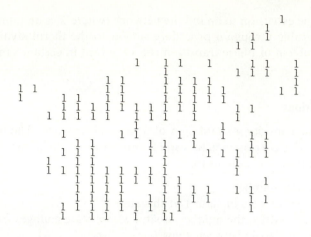

Figure 3.2 Simulation of a renewal network: iteration number = 2; number of 1 = 156; N00 = 701; N01 = 81; N10 = 43; N11 = 75

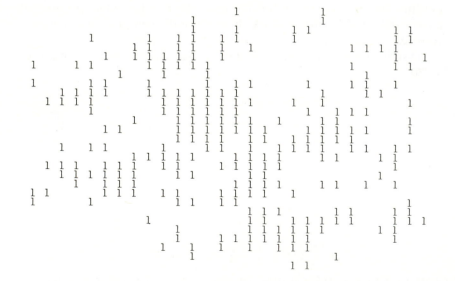

Figure 3.3 Simulation of a renewal network: iteration number = 6; number of 1 = 271; N00 = 535; N01 = 141; N10 = 94; N11 = 130

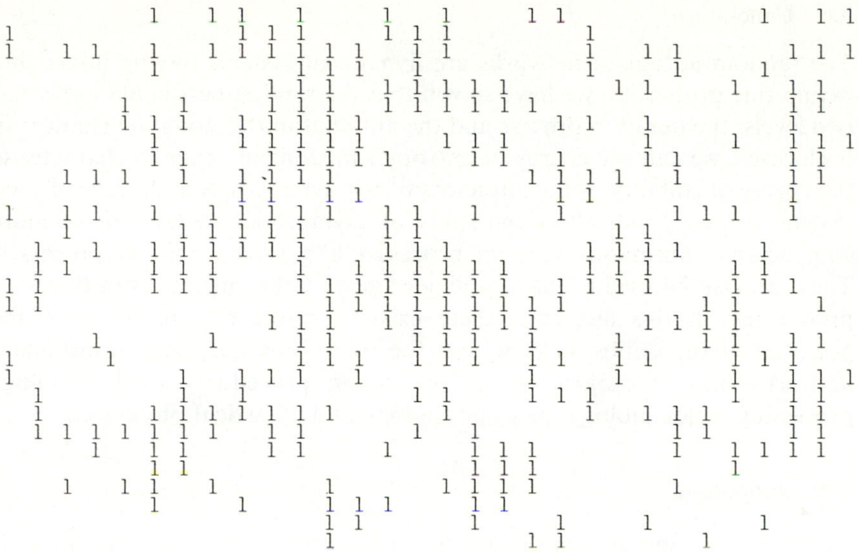

Figure 3.4 Simulation of a renewal network: iteration number = 10; number of
1 = 368; N00 = 368; N01 = 160; N10 = 164; N11 = 208

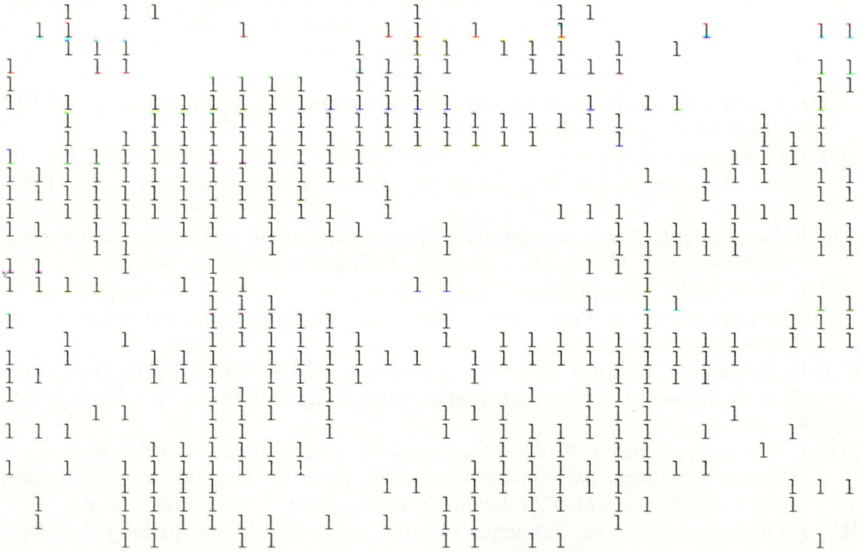

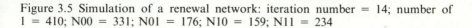

Figure 3.5 Simulation of a renewal network: iteration number = 14; number of
1 = 410; N00 = 331; N01 = 176; N10 = 159; N11 = 234

3.6 Conclusion

The random automata networks are dynamical systems having interesting asymptotic properties; we have shown that the randomness could appear at two levels: the iteration process and the automaton rule for state change. In both cases, we can use energy or entropy functions in order to characterise the degree of stability of the attractors of the dynamics. Simulations of such systems are easy and allow the study of asymptotical behaviour of more complicated continuous random processes like Harris contact processes. They can also be used in the sampling of given Gibbs measures in Bayesian procedures; in this last case, there exists a direct relation between the potential of the Gibbs measure and the rule. Thus the random automata networks have a major role in simulation procedures used in image processing, epidemiology, neurophysiology and statistical physics.

3.7 References

[1] D. J. Amit, Spin glass models for neural networks, in *Proc. Conf. on Physics in Environmental and Biomedical Research, Rome 26–29 Nov. 1985*, (to appear)

[2] M. Arbib, *Theories of Abstract Automata*, Prentice-Hall, Englewood Cliffs, NJ (1969)

[3] J. Besag, Spatial interaction and the statistical analysis of lattice systems, *J. Royal Statist. Soc.*, B36 (1974), 192–326

[4] C. von der Malsburg and E. Bienenstock, Statistical coding and short-term synaptic plasticity: a scheme for knowledge respresentation in brain, in *Disordered systems and Biological Organization*, E. Bienenstock *et al.* (eds), NATO ASI Series, vol. F20, Springer Verlag, New York (1986), 247–72

[5] J. W. Carlyle, *Equivalent Stochastic Sequential Machines*, Elec. Res. Lab. Ser. Berkeley 415, (1961)

[6] K. Chmiel, Cycles in the SC-image of nondeterministic finite automata, ICS-PAS Reports Warsaw 498, (1983).

[7] J. Demongeot, Asymptotic inference for Markov random fields on $\mathbb{Z}^d$, Springer Series in Synergetics 9, Springer Verlag, Berlin, Heidelberg, New York (1981), 254–67

[8] J. Demongeot, *Etude asymptotique d'un processus de contagion*, *Biométrie et Epidémiologie*, J. M. Legay *et al.* (eds), INRA, Paris (1981), 143–52

[9] J. Demongeot, Coupling of Markov processes and Holley's inequalities for Gibbs measures, in *Proc. of the IXth Prague Conference*, Academia, Prague (1983), 183–9

[10] J. Demongeot, Random automata and random fields, in *Dynamical Systems and Cellular Automata*, J. Demongeot *et al.* (eds), Academic Press, New York (1985), 99–110

[11] J. Demongeot, and J. Fricot, Random fields and spatial renewal potentials, in *Disordered Systems and Biological Organization*, E. Bienenstock *et al.* (eds), NATO ASI Series, vol. F20, Springer Verlag, New York (1986), 71–84

[12] J. Demongeot, and M. Tchuente, Cellular automata theory, in *Encyclopedia of Physical Science and Technology*, Academic Press, New York, (to appear)

[13] E. E. Doberkat, *Stochastic Automata: stability, non determinism and prediction*, Springer Verlag, New York (1981)

[14] F. Fogelman-Soulié, Contributions à une théorie du calcul sur réseaux, Thesis, Grenoble (1985)

[15] S. Geman and D. Geman, Stochastic relaxation, Gibbs distributions and the Bayesian restoration of images, *IEEE Trans. Pattern Anal. Machine Intell.*, PAMI-6 (1984), 721–41

[16] S. Goldstein, Entropy increase in dynamical systems, *Israel J. Maths*, 38 (1981), 241–56

[17] U. Grenander, Tutorial in pattern theory, preprint, Brown University, Providence, RI (1983)

[18] X. Guyon, Champs stationnaires sur $\mathbb{Z}^d$: modèles, statistiques et simulations, preprint, Orsay University (1985)

[19] J. J. Hopfield, Neural networks and physical systems with emergent collective computational abilities, *Proc. Nat. Acad. Sci. USA*, 79 (1982), 2554–8

[20] C. von der Malsburg, Algorithms, brain and organization, in *Dynamical Systems and Cellular Automata*, J. Demongeot *et al.* (eds), Academic Press, New York (1985), 235–46

[21] M. Milgram, Thesis, Compiègne (1982)

[22] A. Paz, *Introduction to Probabilistic Automata*, Academic Press, New York (1971)

[23] C. J. Preston, *Gibbs States on Countable Sets*, Cambridge University Press, Cambridge (1974)

[24] M. O. Rabin, Probabilistic automata, *Inf. & Control*, 6 (1966), 230–48

[25] A. Salomaa, On probabilistic automata with one input letter, Amn. Univ. Turku. Ser. A1 (1965), 85

[26] F. Spitzer, Introduction aux processus de Markov à paramètres dans $\mathbb{Z}^d$, Lecture Notes in Maths 390 (1974), 114–89

[27] P. Tautu, A stochastic automaton model for interaction systems, in *Perspectives in Probability and Statistics*, J. Gani (ed.), Applied Prob. Trust, London (1975), 403–15

[28] G. Y. Vichniac, Simulating physics with cellular automata, *Physica*, 10D (1984), 96–116

[29] S. Wolfram, Universality and complexity in cellular automata, *Physica*, 10D (1984), 1–35

Lyapunov functions associated to automata networks

4.1 Introduction

Recently, automata networks have been introduced as a modelling tool in several fields: associative memories, pattern recognition [1, 3, 4, 6, 11, 12], natural languages [16], spin glass problems [1, 5, 9], social dynamics [7, 9], and so on.

One of the most difficult problems has been to find some quantities driving the network's dynamics. Obviously, this is not always possible; e.g. the 'Game of Life' introduced by Conway [2] simulates a universal Turing machine and general neural networks any finite automaton [13]. In this chapter we consider some interesting cases where we can associate global quantities that allow us to determine the dynamic behaviour of the network.

4.2 Threshold networks

Threshold networks were introduced by McCulloch and Pitts [13] to model neural electrical activity. They consist of a network of interconnected elements. Each element i has an internal state (usually -1 or $+1$), and interacts at discrete time steps with other elements through its output lines (identified with its internal state). In turn, it is influenced by other sites through its input lines, and updates its own state x_i according to a threshold function depending only on the input lines, as shown in Fig. 4.1.

where $y = \mathbf{1}(\Sigma_{j=1,n}a_{ij}x_j(t) - b_i)$ and $x_1, \ldots, x_n$ belong to the state set $E = \{-1, +1\}$ and

$$\mathbf{1}(u) = \begin{cases} -1 & \text{if } u < 0 \\ +1 & \text{if } u \geqslant 0 \end{cases}$$

The coefficients a_{ij} are often called *synaptic weights* and b_i the *threshold* associated to element i.

In the general case (without hypotheses on the weights a_{ij}) such a network

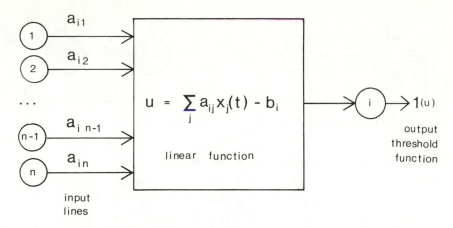

Figure 4.1 Threshold function

simulates any finite automaton [9, 13]. Here we shall suppose that the coefficient matrix $A = (a_{ij})$ is symmetric. This hypothesis is the usual one in cellular arrays and in several applications.

There are two important ways to implement the updating of the states on the network: *parallel* and *sequential*. For the former we have the following iteration scheme:

$$x_i(t + 1) = \mathbf{1}\left(\sum_{j=1}^{n} a_{ij}x_j(t) - b_i \right) \quad t = 0, 1, 2, \ldots; \quad i = 1, \ldots, n$$

and for the sequential mode:

$$x_i(t + 1) = \mathbf{1}\left(\sum_{j<i} a_{ij}x_j(t + 1) + \sum_{j\geq i} a_{ij}x_j(t) - b_i \right)$$

$$t = 0, 1, 2, \ldots; \quad i = 1, \ldots, n$$

that it to say, site i updates its state immediately after site $i - 1$ has done.

Also we shall analyse *block-sequential* iterations that contain as particular cases the parallel and sequential modes.

These iteration modes present different dynamic behaviour but there are similarities. For instance, the fixed points of a threshold network are independent of the iteration mode. That is not the case for more general cycles or transient phenomena.

Example 1 Let us take the one-dimensional torus $\mathbb{Z}_n$ (see Fig. 4.2) such that for each site i we have the following parallel transition function:

$$x_i(t + 1) = \mathbf{1}(x_{i-1}(t) - x_i(t) + x_{i+1}(t)); \, x_i(t) \in \{-1, +1\}$$

where indices are taken modulo n.

Figure 4.2 One-dimensional torus $\mathbb{Z}_n$

For this network, matrix A is the following:

$$A = \begin{array}{c} \\ 0 \\ 1 \\ \\ n-1 \end{array} \begin{pmatrix} \begin{array}{ccccccccc} 0 & 1 & 2 & 3 & 4 & 5 & & n-2 & n-1 \end{array} \\ \begin{array}{ccccccccc} -1 & 1 & 0 & 0 & 0 & 0 & \ldots & 0 & 1 \\ 1 & -1 & 1 & 0 & 0 & 0 & \ldots & 0 & 0 \\ & \vdots & & & & & & & \vdots \\ 1 & 0 & 0 & 0 & 0 & 0 & \ldots & 1 & -1 \end{array} \end{pmatrix}$$

The local transition function can be interpreted in the following way: cell i takes the dominant state on its neighbouring cells $i - 1$ and $i + 1$, and in case of a tie (equal value for both cells in the neighbourhood) cell i changes its previous state. This kind of local behaviour models certain situations in spin glass models as we shall see later.

Some dynamical patterns are shown in Fig. 4.3.

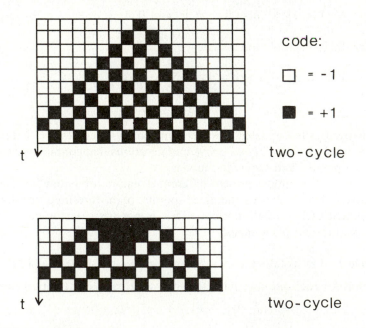

Figure 4.3 Dynamical patterns

Example 2 Now we change the network for that shown in Fig. 4.4 [1, 15],

Figure 4.4 Network for example 2

with transition functions:

$$x_0(t + 1) = \mathbf{1}(x_1(t))$$
$$x_i(t + 1) = \mathbf{1}(x_{i-1}(t) + x_{i+1}(t)) \quad \text{for} \quad 1 \le i \le n - 2$$
$$x_{n-1}(t + 1) = \mathbf{1}(x_{n-2}(t) + x_{n-1}(t))$$

Here matrix A is given by:

$$A = \begin{pmatrix} 0 & 1 & 0 & \dots & 0 & 0 \\ 1 & 0 & 1 & \dots & 0 & 0 \\ & \vdots & & & \vdots & \\ 0 & 0 & 0 & \dots & 1 & 1 \end{pmatrix}$$

For the initial configuration $x(0) = (1, -1, -1, -1)$ the dynamic is shown in Fig. 4.5.

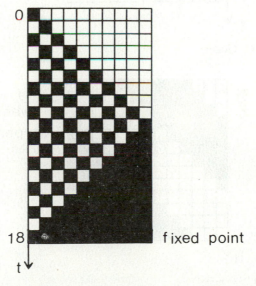

Figure 4.5 Dynamic for example 2

In this case there exists a very long transient phase before one gets the fixed point $(1, 1, \dots, 1)$.

Now, for the two previous networks we shall update states in *sequential mode*, from left to right. For the first example we obtain the local transition functions:

$$x_0(t + 1) = \mathbf{1}(x_{n-1}(t) - x_0(t) + x_1(t))$$
$$x_i(t + 1) = \mathbf{1}(x_{i-1}(t + 1) - x_i(t) + x_{i+1}(t)) \quad \text{for} \quad i \geq 1$$

where the indices are taken modulo n. For the same initial configuration exhibited in the parallel iteration we have the behaviour shown in Fig. 4.6. In this example the symmetry breaking, introduced in the network by the sequential iteration, produces configurations 'walking' in the array as gliders or vehicles. Another example of dynamical behaviour is shown in Fig. 4.7.

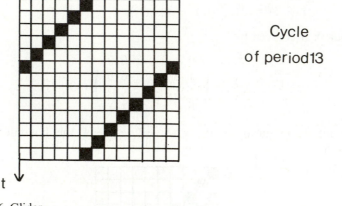

Cycle

of period13

t

Figure 4.6 Glider

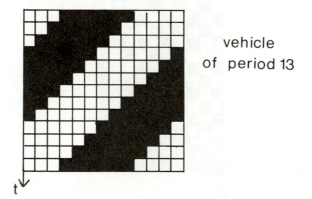

vehicle

of period 13

t

Figure 4.7 Vehicle

If we slightly change the transition rule of one cell in the network, for instance the 6th cell,

$$x_6(t + 1) = \mathbf{1}(x_5(t + 1) + x_6(t) + x_7(t))$$

(i.e., in case of a tie the 6th cell stays in its previous state)

$$(-1, x, +1) \mapsto x \quad \text{and} \quad (+1, x, -1) \mapsto x$$

then for the previous initial configuration $x(0) = (-1, \ldots, -1, 1, -1)$ we obtain a freezing phenomena: any configuration going to a fixed point (Fig. 4.8).

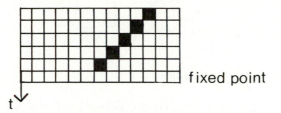

f ixed point

t

Figure 4.8 Fixed point

Let us now take the second example with sequential iteration, (Fig. 4.9):

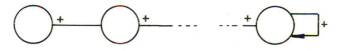

Figure 4.9 Network with sequential iteration

$$x_0(t + 1) = \mathbf{1}[x_1(t)]$$
$$\vdots$$
$$x_i(t + 1) = \mathbf{1}[x_{i-1}(t + 1) + x_{i+1}(t)] \quad \text{for} \quad 1 \le i \le n-2$$
$$\vdots$$
$$x_{n-1}(t + 1) = \mathbf{1}[x_{n-2}(t + 1) + x_{n-1}(t)]$$

The dynamic is that shown in Fig. 4.10. Here the transient is shorter than in the parallel case.

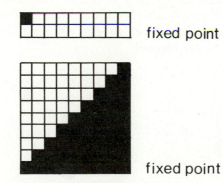

fixed point

fixed point

Figure 4.10 Dynamic of network with sequential iteration

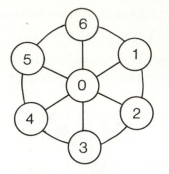

Figure 4.11 Network of example 3

Example 3 Let us now take the network of Fig. 4.11, with transition functions:
For $|V(i)|$ odd:

$$x_i(t + 1) = \begin{cases} -1 & \text{if } \Sigma_{j \in V(i)} x_j(t) < 0 \\ +1 & \text{otherwise} \end{cases}$$

for $|V(i)|$ even:

$$x_i(t + 1) = \begin{cases} -1 & \text{if } \Sigma_{j \in V(i)} x_j(t) < 0 \\ x_i(t) & \text{if } \Sigma_{j \in V(i)} x_j(t) = 0 \\ +1 & \text{otherwise} \end{cases}$$

where $V(i)$ is the neighbourhood set associated to site i:

$$V(i) = \{j \in \{1, \ldots, n\}; a_{ij} \neq 0\}$$

Here we have a model of group dynamics. We can see the network as a society where individuals are interrelated with friendly, hostile or indifferent feelings ($a_{ij} > 0$, $a_{ij} < 0$ and $a_{ij} = 0$). The set $\{-1, +1\}$ corresponds to two opinions, or votes, that individuals must express at all times. In the parallel case we have, for instance, the dynamic of Fig. 4.12. In this diagram the central node can be interpreted as a 'president' that stays constant for any time while the population is always divided in a tie of opinions. With the sequential mode the president configuration goes to a fixed point where all individuals have the same opinion.

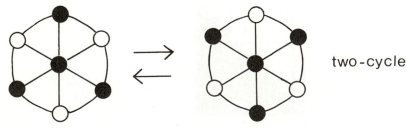

Figure 4.12 Group dynamics

4.3 Dynamical behaviour of the parallel iteration

Let us take the general threshold network with transition functions:

$$x_i(t + 1) = \mathbf{1}\left(\sum_{j=1}^{n} a_{ij} x_j(t) - b_i\right) \quad t = 0, 1, 2, \ldots; i = 1, \ldots, n$$

where $(x_1(t), \ldots, x_n(t)) \in \{-1, +1\}^n$, A is a real $n \times n$ symmetric matrix, and $b = (b_1, \ldots, b_n)$ is an n-dimensional real threshold vector. We shall analyse the dynamic behaviour of this model by associating to the network a monotone operator, or *Lyapunov function*, depending only on the initial configuration, matrix A and threshold vector b. This tool will allow us to characterise the dynamic.

4.3.1 *Lyapunov function for parallel iteration*

Without loss of generality we shall suppose that the threshold functions are *strict*; that is, for any vector $y \in \{-1, +1\}^n$ and for any $i \in \{1, \ldots, n\}$:

$$\sum_{j=1}^{n} a_{ij} y_j - b_i \neq 0$$

In fact, for an arbitrary threshold function we can always find, without changing the function, small perturbations on the thresholds such that the previous summation never vanishes. For instance:

x_1	x_2	f	g
-1	-1	-1	-1
-1	$+1$	$+1$	$+1$
$+1$	-1	$+1$	$+1$
$+1$	$+1$	$+1$	$+1$

$$f(x_1, x_2) = \mathbf{1}(x_1 + x_2)$$

$$g(x_1, x_2) = \mathbf{1}(x_1 + x_2 + 1)$$

$$f = g$$

Let us denote $\Sigma_1 = x_1 + x_2$ and $\Sigma_2 = x_1 + x_2 + 1$, hence Fig. 4.13.

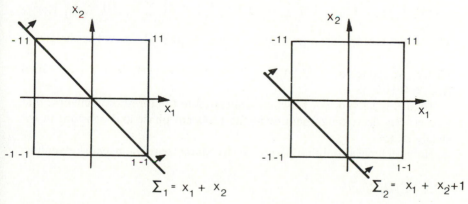

Figure 4.13

$\Sigma_1(-1, +1) = \Sigma_1(1, -1) = 0$

while $\Sigma_2(x_1, x_2) \neq 0$ for any $x_1, x_2 \in \{-1, +1\}$

Let us now consider the following bilinear operator:

$$E(x(t+1), x(t)) = -\langle x(t+1), Ax(t) \rangle + \langle b, x(t+1) + x(t) \rangle$$

where $\{x(t); t \geq 0\}$ is the trajectory of the automata from the initial condition $x(0) \in \{-1, +1\}^n$ and $\langle, \rangle$ denotes the usual inner product in $\mathbb{R}^n$, that is:

$$\langle u, v \rangle = \sum_{i=1}^{n} u_i v_i$$

Since A is a symmetric matrix we have

$$
\begin{aligned}
\Delta_t E &= E(x(t+1), x(t)) - E(x(t), x(t-1)) \\
&= -\langle x(t+1) - x(t), Ax(t) - b \rangle \\
&= -\sum_{i=1}^{n} \left(x_i(t+1) - x_i(t-1) \right) \left(\sum_{j=1}^{n} a_{ij} x_j(t) - b_i \right)
\end{aligned}
$$

For each term of this summation we have:

$$x_i(t+1) = x_i(t-1) \Rightarrow (x_i(t+1) - x_i(t-1)) \left(\sum_{j=1}^{n} a_{ij} x_j(t) - b_i \right) = 0$$

$$x_i(t+1) \neq x_i(t-1) \Rightarrow (x_i(t+1) - x_i(t-1)) \left(\sum_{j=1}^{n} a_{ij} x_j(t) - b_i \right) > 0$$

The first property is obvious, and for the second let us suppose for instance that $x_i(t+1) = +1$ and $x_i(t-1) = -1$. Clearly, from the definition of the strict threshold function we have $\sum_{j=1}^{n} a_{ij} x_j(t) - b_i > 0$ and thus:

$$(x_i(t+1) - x_i(t-1)) \left(\sum_{j=1}^{n} a_{ij} x_j(t) - b_i \right) = 2 \left(\sum_{j=1}^{n} a_{ij} x_j(t) - b_i \right) > 0$$

The other case, that is, $x_i(t+1) = -1$ and $x_i(t-1) = +1$ is similar.

Hence we conclude that $\Delta_t E \leq 0$ and if $x(t+1) \neq x(t-1)$, then $\Delta_t E < 0$.

Then E is a Lyapunov function associated to the threshold network, that is, a strictly decreasing operator in the transient phase and constant in the steady state.

Now we have a tool that permits us to characterise the network dynamic.

4.3.2 Cycle length

Since the set $\{-1, +1\}^n$ is finite, for any initial vector we finally get to a stationary regime. Let us suppose that we obtain a cycle of period T; that is:

$$x(0) \rightarrow x(1) \rightarrow \ldots \rightarrow x(T-1) \rightarrow x(0) \rightarrow \ldots$$

where $x_i(t+1) = \mathbf{1}\left(\sum_{j=1}^{n} a_{ij}x_j(t) - b_i\right)$ for any $i = 1, \ldots, n$ and $t = 0, \ldots,$
$T - 1$ and the indices, t, are taken modulo T.

Since $x(t) \neq x(t')$ for any $t \neq t'$ and $t, t' \in \{0, \ldots, T-1\}$ we obtain:

$$E(x(2), x(1)) - E(x(1), x(0)) < 0$$
$$E(x(3), x(2)) - E(x(2), x(1)) < 0$$
$$\vdots$$
$$E(x(T-1), x(T-2)) - E(x(T-2), x(T-3)) < 0$$
$$E(x(0), x(T-1)) - E(x(T-1), x(T-2)) < 0$$
$$E(x(1), x(0)) - E(x(0), x(T-1)) < 0$$

hence:

$$E(x(1), x(0)) > E(x(2), x(1)) > \ldots > E(x(0), x(T-1)) > E(x(1), x(0))$$

which is a contradiction. Then the only possibilities are $x(t+1) = x(t)$ or $x(t+1) = x(t-1)$, that is to say in a threshold network with parallel iteration the periods of cycles are 1 or 2 and in these cases $\Delta_t E = 0$. Hence we conclude that:

> If A is a symmetric $n \times n$ real matrix, the cycles of the parallel iteration on a strict threshold network have periods 1 (fixed points) or 2.

4.3.3 *Transient length*

With the same tool, the Lyapunov function E, we can analyse the convergence speed, or transient length, for the parallel update.
Since $\{-1, +1\}^n$ is a finite set, we can bound the Lyapunov function E in the following way: let $\{x(t)/t = 0, \ldots, q-1\}$ be a transient trajectory; that is,

$$x(0) \rightarrow x(1) \rightarrow \ldots \rightarrow x(q-1) \rightarrow x(q) \rightarrow \ldots$$

where $x(q)$ is the first vector belonging to the steady state associated to the initial configuration $x(0)$. Since in the transient the function E is strictly decreasing there exists:

$$e = \min_{1 \leq t \leq q-1} |\Delta_t E| = \min_{1 \leq t \leq q-1} |E(x(t+1), x(t)) - E(x(t), x(t-1))| > 0$$

and since $\sum_{j=1}^{n} a_{ij}u_j - b_i \neq 0$ for any vector $u \in \{-1, +1\}^n$, we can define positive quantities:

$$e_i = \min\left\{\left|\sum_{j=1}^{n} a_{ij}u_j - b_i\right|; u \in \{-1, +1\}^n\right\} \quad i = 1, \ldots, n$$

Thus we obtain:

$$-x_i(t + 1)\left(\sum_{j=1}^{n} a_{ij}x_j(t) - b_i\right) + x_i(t)b_i \le -e_i + |b_i| \quad \text{for} \quad i = 1, \ldots, n$$

Hence, from the definition of E:

$$E(x(t + 1), x(t)) \le -\sum_{i=1}^{n} e_i + \|b\|$$

On the other hand it is easy to see that:

$$E(x(t + 1), x(t)) \ge -\|A\| - 2\|b\|$$

where $\|A\| = \Sigma_{i,j}|a_{ij}|$ and $\|b\| = \Sigma_i|b_i|$.

Then we can bound E in the following way:

$$-\|A\| - 2\|b\| \le E(x(t + 1), x(t)) \le -\sum_{i=1}^{n} e_i + \|b\|$$

Furthermore, since in the transient phase the Lyapunov function E is strictly decreasing:

$$E(x(t + 1), x(t)) \le E(x(t), x(t - 1)) - e \le E(x(1), x(0)) - te$$

Hence for $t = q$:

$$-\|A\| - 2\|b\| \le E(x(q + 1), x(q)) \le E(x(1), x(0)) - qe$$

From previous equations we conclude:

$$qe \le E(x(1), x(0)) + \|A\| + 2\|b\|$$

Hence the transient of the threshold network is bounded by:

$$q \le \left(-\sum_{i=1}^{n} e_i + 3\|b\| + \|A\|\right)\bigg/e$$

Let us now consider a particular case where we explicitly find the parameters e and e_i:

$$x_i(t + 1) = 1\left(\sum_{j=1}^{n} a_{ij}x_j(t)\right) \quad \text{for} \quad i = 1, \ldots, n \text{ and } a_{ij} \in \mathbb{Z}$$

It is easy to see that if $\sum_{j=1}^{n} a_{ij}$ is odd, then for any vector $u \in \{-1, +1\}^n$:

$$\sum_{j=1}^{n} a_{ij}u_j(t) \ne 0$$

In the other case, i.e when $\sum_{j=1}^{n} a_{ij}$ is an even number, we can always take a threshold $b_i = 1$ such that:

$$\sum_{j=1}^{n} a_{ij} u_j(t) + b_i \neq 0$$

Then we have the following parallel network equivalent to the previous one:

$$x_i(t + 1) = \mathbf{1}\left(\sum_{j=1}^{n} a_{ij} x_j(t) + b_i\right) \quad \text{for } i = 1, \ldots, n$$

$$\text{with } b_i = \begin{cases} 0 & \text{if } \sum_{j=1}^{n} a_{ij} \text{ is odd} \\ 1 & \text{if } \sum_{j=1}^{n} a_{ij} \text{ is even} \end{cases}$$

On the other hand, since the coefficients a_{ij} are integers, we have $e = 2$ and $e_i \geq 1$ for $i = 1, \ldots, n$. From the previous general bound on the transient length we obtain:

$$q \leq (\|A\| - n + 3\alpha)/2$$

where $\alpha = \sum_{i=1}^{n} b_i =$ the number of even lines in matrix A.

This bound is the best because it is attained in the network of example 2 (see Fig. 4.14) with the initial configuration $1(-1)^{n-1}$ and parallel updating.

Figure 4.14 Network of example 2

In the general case, that is, matrix A not necessarily with integral entries, we can find a family of symmetric threshold networks with an extremely slow convergence speed, $0(2^{n/3})$, [9] but in most of the cases the transient is very short, usually linear in n, the number of cells in the network. For instance Fig. 4.15 shows the 2-dimensional torus with von Neumann neighbourhood.

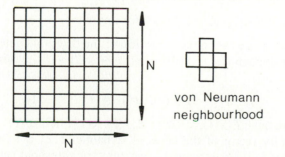

Figure 4.15 2-dimensional torus

Where $n = N^2$ is the number of cells and the local functions are:

$$1\left(\sum_{j \in V(i)} x_j(t)\right)$$

We have $b_i = 0$ and $e_i = 1$ for any i, $e = 2$ and $\|A\| = 5N^2 = 5n$, hence $q \leqslant 2n$.

4.3.4 *Discrete optimisation problem and stationary configurations*

Let us assume that $b_i = 0$ for any $i = 1, \ldots, n$. In this particular case the Lyapunov function associated to the network is:

$$E(x(t + 1), x(t)) = -\langle x(t + 1), Ax(t) \rangle \quad t \geq 0$$

Since

$$x_i(t + 1) = 1\left(\sum_{j=1}^{n} a_{ij} x_j(t)\right)$$

we have

$$x_i(t + 1) \sum_{j=1}^{n} a_{ij} x_j(t) = \left|\sum_{j=1}^{n} a_{ij} x_j(t)\right|$$

and hence

$$E(x(t + 1), x(t)) = -\sum_{i=1}^{n}\left|\sum_{j=1}^{n} a_{ij} x_j(t)\right| = -\|Ax(t)\|_1$$

where $\|u\|_1 = \sum_{i=1}^{n} |u_i|$ is the Manhattan norm in $\mathbb{R}^n$. Since E is a non-increasing function, from the previous equality we conclude that $\|Au\|_1$ is a Lyapunov function for the network dynamics.

Let us consider now the following discrete optimisation problem:

$$\min \{-\|Av\|_1; v \in \{-1, +1\}^n\} \tag{4.1}$$

and let C be the set of global optima of (4.1) and $u \in C$, that is

$$-\|Au\|_1 \leq -\|Av\|_1 \quad \text{for any } v \in \{-1, +1\}^n$$

If we take $x(0) = u$ as an initial condition of the threshold network we have the following dynamic: $u = x(0) \to x(1) \to x(2) \to \ldots$, with decreasing energies:

$$-\|Au\|_1 = E(x(1), x(0)) \geq E(x(2), x(1))$$

Since $u \in C$ we have $E(x(1), x(0)) = E(x(2), x(1))$, hence $x(2) = x(0)$; thus u is a stationary vector of the network dynamic. That is to say, the fixed points or the two-cycles associated to a symmetric threshold network include as a subset the optimal vectors of the discrete optimisation problem (4.1).

We can then interpret the global behaviour of the parallel iteration on threshold functions as a hill-climbing algorithm for the problem (4.1). Stationary points are local minima of the function $\| Av \|_1$ on the hypercube $\{-1, +1\}^n$.

4.4 Sequential iteration

Let us take now the sequential update on a symmetric strict threshold network:

$$x_i(t + 1) = \mathbf{1}\left(\sum_{j=1}^{i-1} a_{ij}x_j(t + 1) + \sum_{j=i}^{n} a_{ij}x_j(t) - b_i \right) \quad \text{for } i = 1, \ldots, n$$

and the following function:

$$E(x(t)) = -\langle x(t), Ax(t)\rangle/2 + \langle b, x(t)\rangle$$

The sequential iteration at time t can be decomposed into succesive steps, where at time $t + i/n$ only element i changes state, that is to say:

$$x(t) = (x_1(t), \ldots, x_n(t))$$

$$x(t + k/n) = (x_1(t + 1), \ldots, x_k(t + 1), x_{k+1}(t), \ldots, x_n(t)) \quad 1 < k \le n$$

hence

$$\Delta_t E = E(x(t + 1)) - E(x(t)) = \sum_{k=1}^{n} \{E(x(t + k/n)) - E(x(t + (k - 1)/n))\}$$

and, since A is symmetric:

$$d_k E = E(x(t + k/n)) - E(x(t + (k - 1)/n))$$

$$= -[x_k(t + 1) - x_k(t)]\left[\sum_{j=1}^{k-1} a_{kj}x_j(t + 1) + \sum_{j=k}^{n} a_{kj}x_j(t) - b_k \right]$$

$$-[x_k(t + 1) - x_k(t)]a_{kk}x_k(t) - x_k(t + 1)a_{kk}x_k(t + 1)/2$$

$$+ x_k(t)a_{kk}x_k(t)/2$$

$$= -[x_k(t + 1) - x_k(t)]\left[\sum_{j=1}^{k-1} a_{kj}x_j(t + 1) + \sum_{j=k}^{n} a_{kj}x_j(t) - b_k \right]$$

$$- a_{kk}[x_k(t + 1) - x_k(t)]^2/2$$

From the definition of strict threshold functions and sequential iteration, the first term in the previous equation is strictly negative. On the other hand, if we assume that diag $A = (a_{11}, \ldots, a_{nn}) \ge (0, \ldots, 0)$, that is, any diagonal entry of matrix A is nonnegative, then the second term of the previous equality is negative. We thus conclude:

If A is a symmetric matrix with nonnegative diagonal entries then:

$$x(t + 1) \neq x(t) \Rightarrow \Delta_t E < 0$$

That is to say, E is a Lyapunov function associated to the sequential iteration on the threshold network.

Furthermore, the previous equation implies that in the steady state *there exist only fixed points*, in fact if we have a cycle of period greater than 2:

$$x(0) \to x(1) \to \ldots \to x(T - 1) \to x(0) \to \ldots \Rightarrow x(t) \neq x(t') \text{ for any } t \neq t'$$

hence $E(x(0)) > E(x(1)) > \ldots > E(x(T - 1)) > E(x(0))$, which is a contradiction.

Obviously, if diag A is not nonnegative we can find cycles. It is sufficient to take, for instance, a_{11} sufficiently negative so that the cell 1 has a two-cycle.

Negativity on diagonal entries can be interpreted in the framework of the spin glass problem [1, 9].

Similarly to the parallel case, stationary vectors of the sequential iteration are local minima of a discrete quadratic optimisation problem:

$$\min \{ - \langle v, Av \rangle /2 + \langle b, v \rangle; v \in \{-1, +1\}^n \} \tag{4.2}$$

If u is a global minimum of (4.2) it is easy to see, under the previous hypotheses, that u is a fixed point of the sequential iteration on the threshold network. In fact, taking $x(0) = u$ we have the following dynamic:

$$u = x(0) \to x(1) \to x(2) \to \ldots$$

Since E is a Lyapunov function: $E(x(0)) \geq E(x(1))$ and since $u = x(0)$ is a global minimum:

$$E(x(0)) = - \langle u, Au \rangle + \langle b, u \rangle \leq - \langle x(1), Ax(1) \rangle + \langle b, x(1) \rangle$$

hence $E(x(0)) = E(x(1))$; that is, $\Delta_t E = \sum_{k=1}^{n} d_k E = 0$. Since for any k, $d_k E \leq 0$ then each term vanishes and, from the definition of $d_k E$, we conclude that $x(1) = x(0)$.

Then $x(0) = u$ is a fixed point of the network.

Example 4   We take the threshold vector $b = (0, \ldots, 0)$ and the following matrix A:

$$A = \begin{pmatrix} 1 & 1 & 0 & 0 & 1 \\ 1 & 1 & 1 & 0 & 0 \\ 0 & 1 & 1 & 1 & 0 \\ 0 & 0 & 1 & 1 & 1 \\ 1 & 0 & 0 & 1 & 1 \end{pmatrix}$$

Clearly diag $A \geq 0$. Furthermore, since each line has an odd sum of entries we get the strict hypothesis. The fixed points of this network are the following:

11111; 111 − 1 − 1; − 1111 − 1; − 1 − 1111; 1 − 1 − 111; 11 − 1 − 11;
11 − 1 − 1 − 1; − 111 − 1 − 1; − 1 − 111 − 1; − 1 − 1 − 11; 1 − 1 − 1 − 11;
− 1 − 1 − 1 − 1 − 1

with energies:

$E(11111) = E(-1-1-1-1-1) = -7.5$ and for all the others the
energy is -3.5. Hence the global minima are the vectors (11111) and
$(-1-1-1-1-1)$.

4.5 Modelling spin glasses by threshold networks

We will now consider a particular case of sequential iteration on threshold
functions which has been introduced in solid state physics [1, 9]. Some
magnetic alloys are obtained by diluting magnetic impurities in a nonmagne-
tic metal. These systems are called *spin glasses*. Their magnetic properties can
be deduced from the study of a Hamiltonian:

$$H(S) = -\sum_{i,j} J_{ij} S_i S_j - \sum_i F_i S_i$$

where S is the spins vector, S_i is the spin of impurity i, J_{ij} is the interaction
between impurities i and j and F is an external magnetic field. Furthermore,
spin S_i can only take values in a finite set, usually $\{-1, +1\}$.

The ground states of a spin glass are associated to minima of the
Hamiltonian H. Since spins take values in a finite set, finding these minima is
usually a very hard combinatorial problem. More precisely, it has been
shown that, in the two-dimensional problem (i.e., the impurities are supposed
to be distributed regularly at the nodes of a grid, commonly toric) without
magnetic field, ground states can be found in polynomial time; the case with
magnetic field and the three-dimensional case are NP-complete. Hence
physicists can only find exact solutions of this problem in particular cases.
For more general situations, they use hill-climbing algorithms or Monte
Carlo iterations in order to estimate ground states. Formally, this last
method consists in updating a given spins configuration by choosing one spin
at random and eventually changing its state with a probability depending on
a local majority or threshold rule.

We will study sequential iterations of *deterministic* rules, which correspond in the spin glass context to a zero temperature: the choice of the spin
that is to change states is not random, but fixed by a given permutation of the
sites in the graph which we will assume, without loss of generality, to be the
identity.

Let $n \in \mathbb{N}$, (I_0, I_1) be a partition of $\{1, \ldots, n\}$. A *generalised spin glass* is a
threshold network whose components satisfy:

For any $i \in I_0$:

$$x_i(t + 1) = \begin{cases} -1 & \text{if } \Sigma_{j<i}a_{ij}x_j(t + 1) + \Sigma_{i<j\leq n}a_{ij}x_j(t) - b_i < 0 \\ x_i(t) & \text{if } \Sigma_{j<i}a_{ij}x_j(t + 1) + \Sigma_{i<j\leq n}a_{ij}x_j(t) - b_i = 0 \\ +1 & \text{if } \Sigma_{j<i}a_{ij}x_j(t + 1) + \Sigma_{i<j\leq n}a_{ij}x_j(t) - b_i > 0 \end{cases}$$

For any $i \in I_1$:

$$x_i(t + 1) = \begin{cases} -1 & \text{if } \Sigma_{j<i}a_{ij}x_j(t + 1) + \Sigma_{i<j\leq n}a_{ij}x_j(t) - b_i < 0 \\ -x_i(t) & \text{if } \Sigma_{j<i}a_{ij}x_j(t + 1) + \Sigma_{i<j\leq n}a_{ij}x_j(t) - b_i = 0 \\ +1 & \text{if } \Sigma_{j<i}a_{ij}x_j(t + 1) + \Sigma_{i<j\leq n}a_{j}x_j(t) - b_i > 0 \end{cases}$$

where A is a symmetric $n \times n$ matrix with diag $A = (0, \ldots, 0)$. Note that the threshold functions are not strict.

We can interpret this model in the following way: in case of a tie, the ith site only changes its state if i belongs to the set I_1. Intuitively this situation should originate unstable behaviour (long cycles). On the other hand, the fact that a site belongs to set I_0 should freeze its dynamic behaviour (convergence to fixed points).

Let us take the energy associated to the model:

$$E(v) = -\langle v, Av \rangle/2 + \langle b, v \rangle \quad v \in \{-1, +1\}^n$$

Similarly to the preceding analysis of the sequential iteration, since A is a symmetric matrix we have:

$$d_k E = E(x(t + k/n)) - E(x(t + (k - 1)/n))$$

$$= -(x_k(t + 1) - x_k(t))\left(\sum_{j<k} a_{kj}x_j(t + 1) + \sum_{j>k} a_{kj}x_j(t) - b_k \right)$$

thus if $k \in I_0$:

$$x_k(t + 1) \neq x_k(t) \Rightarrow \sum_{j<k} a_{kj}x_j(t + 1) + \sum_{j>k} a_{kj}x_j(t) - b_k \neq 0$$

and furthermore:

$$x_k(t + 1) = 1 \Rightarrow \sum_{j<k} a_{kj}x_j(t + 1) + \sum_{j>k} a_{kj}x_j(t) - b_k > 0$$

$$x_k(t + 1) = -1 \Rightarrow \sum_{j<k} a_{kj}x_j(t + 1) + \sum_{j>k} a_{kj}x_j(t) - b_k < 0$$

Then we conclude:

$$k \in I_0 \quad \text{and} \quad x_k(t + 1) \neq x_k(t) \Rightarrow d_k E < 0$$

if $k \in I_1$, $x_k(t + 1) \neq x_k(t)$ implies only $d_k E \leq 0$, that is, not necessarily strictly negative.

We thus conclude that, for this kind of network, E is decreasing, but not necessarily strictly, in the transient phase. That is to say, we can have two successive different configurations with $\Delta_t E = 0$. Furthermore, in the steady state we might have cycles where, for any couple of successive steps, $\Delta_t E = 0$.

More precisely, from previous analysis we have:

For any $i \in I_0$ there exists $t^* \in \mathbb{N}$ such that for any $t \geq t^*$, $x_i(t) = x_i(t^*)$; i.e., the ith site becomes stable.

In the steady state, the only possible oscillating sites belong to I_1 and each site changes state only in the case of a tie (otherwise the energy is strictly decreasing).

Clearly, elements in I_0 help to freeze the network in fixed points: neighbours of such stable elements tend also to be stable. Furthermore, oscillating elements (sites belonging to I_1) are scarce since, in the steady state, the combinatorics needed for this situation makes it very rare: in the steady state a tie must happen at each step when an element in I_1 is supposed to change state. Simulations usually showed less than 20% of the sites in I_1 oscillating. Roughly speaking we could say that the periods of cycles are very small by comparison with the 2^n possible configurations of the network.

In Section 4.2 we examined a one-dimensional model of the previous situation (example 1 with sequential updating). There, we saw that when $I_0 = \varnothing$ there exist cycles (gliders or vehicles) with periods $0(n)$ and, if $I_0 \neq \varnothing$, the network is frozen (only fixed points in the steady state).

Example 5 Let us take the 3×3 two-dimensional torus with $I_0 = \varnothing$ and the following sequential iteration mode:

$$x_{ij}(t + 1)$$
$$= \begin{cases} -1 & \text{if} \quad -x_{i-1j}(t + 1) - x_{ij-1}(t + 1) - x_{ij+1}(t) - x_{i+1j}(t) < 0 \\ -x_i(t) & \text{if} \quad -x_{i-1j}(t + 1) - x_{ij-1}(t + 1) - x_{ij+1}(t) - x_{i+1j}(t) = 0 \\ +1 & \text{if} \quad -x_{i-1j}(t + 1) - x_{ij-1}(t + 1) - x_{ij+1}(t) - x_{i+1j}(t) > 0 \end{cases}$$

for any $i, j = 0, 1, 2$ where indices i, j are taken modulo 3.

In this case we can obtain cycles such as that in Fig. 4.16.

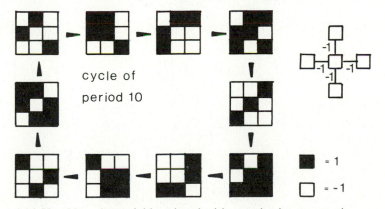

Figure 4.16 Von Neumann neighbourhood with negative interconnections

4.6 Block-sequential iteration

An *ordered partition* of the set of sites $\{1, \ldots, n\}$ is a partition $\{I_k; k = 1, \ldots, p\}$ such that:

$$\forall s \in I_i, \forall s' \in I_j \quad i < j \Rightarrow s < s'$$

The *block-sequential iteration* associated to the ordered partition $\{I_k\}$ is defined by:
$$\forall k \in \{1, \ldots, p\}, \forall i \in I_k$$

$$x_i(t + 1) = 1\left(\sum_{i \in A_k} a_{ij}x_j(t + 1) + \sum_{i \in B_k} a_{ij}x_j(t) - b_i\right)$$

where $\quad A_k = I_1 \cup \ldots \cup I_{k-1} \quad$ and $\quad B_k = I_k \cup \ldots \cup I_p$

Two important cases are the sequential and parallel iterations, which we have previously studied. The former corresponds to the partition $\{\{1\}, \{2\}, \ldots, \{n\}\}$ and the latter to the trivial partition $\{1, \ldots, n\}$.

As in Section 4.5, let E be the following functional:

$$E(v) = -\langle v, Av\rangle/2 + \langle b, v\rangle$$

Also, let us suppose that the matrices associated with each partition set I_k – i.e. matrices $A(k) = (a_{ij})$, where $i, j \in I_k$ – are nonnegative definite on the set $\{-2, 0, 2\}$ for any $k = 1, \ldots, p$.

Let us decompose the block sequential iteration at time t into p successive steps, where at time $t + k/p$, $k = 1, \ldots, p$, elements in block I_k only can change state.

Let $\Delta_t E = E(x(t + 1)) - E(x(t))$

$$= \sum_{k=1}^{p} [E(x(t + k/p)) - E(x(t + (k - 1)/p))]$$

Since A is a symmetric matrix, we have:

$$\Delta_t E = -\sum_{k=1}^{p} \sum_{i \in I_k} [x_i(t + 1) - x_i(t)]\left[\sum_{j \in A_k} a_{ij}x_j(t + 1) + \sum_{j \in B_k} a_{ij}x_j(t) - b_i\right] +$$

$$+ \left[\sum_{i \in I_k} x_i(t + 1) \sum_{j \in I_k} a_{ij}x_j(t + 1)\right]\bigg/2 - \left[\sum_{i \in I_k} x_i(t) \sum_{j \in I_k} a_{ij}x_j(t)\right]\bigg/2$$

Let us denote:

$$d(1, k) = \sum_{i \in I_k} [x_i(t + 1) - x_i(t)] \cdot \left[\sum_{j \in A_k} a_{ij}x_j(t + 1) + \right.$$

$$\left. \sum_{j \in I_k} a_{ij}x_j(t) + \sum_{j \in B_k} a_{ij}x_j(t) - b_i\right]$$

$$d(2, k) = -\sum_{i \in I_k} [x_i(t + 1) - x_i(t)] \cdot \left(\sum_{j \in I_k} a_{ij}[x_j(t + 1) - x_j(t)]\right)\bigg/2$$

Then

$$\Delta_t E = \sum_{k=1}^{p} [d(1, k) + d(2, k)]$$

Since we assumed that the threshold functions are strict, it follows that: $x(t + 1) \neq x(t) \Rightarrow d(1, k) \leq 0$ and there exists at least one $k \in \{1, \ldots, p\}$ such that this quantity is strictly negative.

Moreover, since $x_i(t + 1) - x_i(t) \in \{-2, 0, 2\}$ the assumption on each matrix block $A(k)$ implies $d(2, k) \leq 0$. Hence we conclude $\Delta_t E < 0$. That is to say, the functional E is a Lyapunov function for the block sequential iteration and, similarly to the sequential case, we conclude that:

If A is a symmetric matrix and $A(k)$ is nonnegative definite on the set $\{-2, 0, 2\}$ then the block sequential iteration has only fixed points.

Remark The diagonal dominance assumption on matrices $A(k)$:

$$\forall i \in I_k, a_{ii} \geq \sum_{\substack{i \in I_k \\ j \neq i}} |a_{ij}|$$

is a sufficient condition for the nonnegative definite property.

Furthermore, this hypothesis contains, as a particular case: diag $A \geq 0$, which was our assumption for the sequential iteration (it is sufficient to remark that in this case $I_k = \{k\}$).

Also, as a corollary, we have that, if A is nonnegative definite, the parallel itcration has only fixed points (no two-cycles). In order to see this it is sufficient to apply the previous result to the trivial partition $I_1 = \{1, \ldots, n\}$.

Now, we introduce the following partial order between ordered partitions which will allow us to compare the corresponding block sequential iterations:

Let $P = \{I_k; k = 1, \ldots, p\}$ and $P' = \{I'_k; k = 1, \ldots, q\}$ be two ordered partitions of the set $\{1, \ldots, n\}$. Then P' is *finer* than P, denoted $P' \leq P$, iff any block I_k is the reunion of some blocks I'_j. For instance:

$$\{1, 2\} \{3, 4, 5\} \{6\} \{7, 8, 9\} \leq \{1, 2\} \{3, 4, 5, 6,\} \{7, 8, 9\}$$
$$\leq \{1, 2, 3, 4, 5, 6\} \{7, 8, 9\} \leq \{1, 2, 3, 4, 5, 6, 7, 8, 9\}$$

Endowed with this relation, the set of ordered partitions is a partially ordered set which is a lattice. Now we can state the following result:

Let P be an ordered partition, then if P satisfies the assumptions of nonnegative definiteness of matrices $A(k)$, any partition P' finer than P also satisfies them. This means that, as soon as one partition P satisfies the required assumptions, any $P' \leq P$ has E as a Lyapunov function and thus only fixed points (those of P).

Example 6

$$A = \begin{pmatrix} 1 & 1 & -1 & 1 \\ 1 & 2 & 1 & -2 \\ -1 & 1 & 1 & 1 \\ 1 & -2 & 1 & 2 \end{pmatrix} \quad b = (1, 1, 1, 1)$$

The ordered partition lattice is that shown in Fig. 4.17. Circled ordered partitions satisfy the assumptions of nonnegative definiteness. The iteration graphs of the different block sequential modes associated to the previous lattice are those in Fig. 4.18.

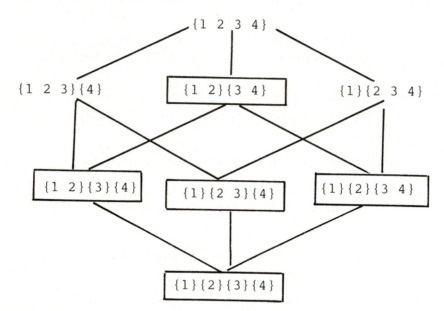

Figure 4.17 Ordered partition lattice of example 6

4.7   Antisymmetric networks

In previous paragraphs we have only studied symmetric networks, that is, networks where A is a symmetric matrix. Here we shall associate a Lyapunov function to an antisymmetric threshold network. Let us suppose that the network is defined by:

$$x_i(t + 1) = \mathbf{1}\left[\sum_{j=1}^{n} a_{ij}x_j(t)\right] \quad \text{for } i = 1, \ldots, n$$

where $x_j(t) \in \{-1, +1\}$ and $A = (a_{ij})$ is an *integer antisymmetric* matrix, that is, $a_{ij} = -a_{ji}$ and $a_{ij} \in \mathbb{Z}$

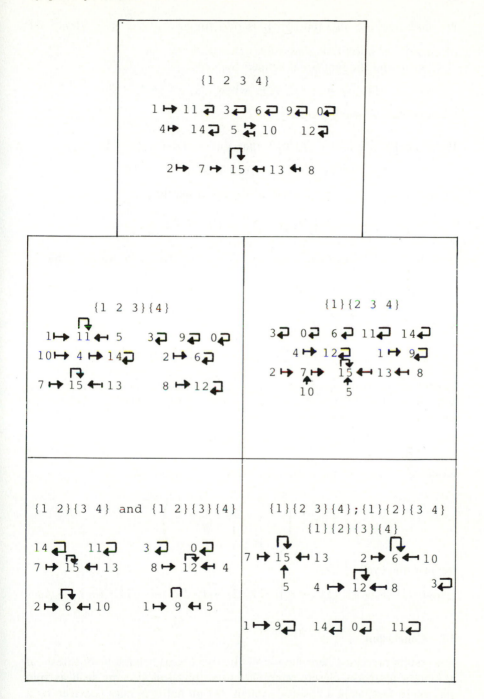

Figure 4.18 Iteration graphs

We shall suppose also that $\sum\limits_{j=1}^{n} a_{ij}$ is odd for any $1 \le i \le n$. Hence the threshold transition functions are strict.

Considering the functional defined by:

$$E(u, v) = -\langle u, Av \rangle \quad \text{where} \quad u, v \in \{-1, +1\}^n$$

we have the following property for E:

If there exists $k \in \{1, \ldots, n\}$ such that $x_k(t + 1) = x_k(t - 1)$ then:

$$\Delta_t E = E(x(t + 1), x(t)) - E(x(t), x(t - 1)) < 0$$

In fact, since A is antisymmetric, it is easy to see that

$$\Delta_t E = -\sum_{i=1}^{n} [x_i(t + 1) + x_i(t - 1)] \sum_{j=1}^{n} a_{ij} x_j(t)$$

since $\Sigma a_{ij} u_j \ne 0$ for any $u \in \{-1, +1\}^n$ and $i = 1, \ldots, n$, we have that

$$-[x_i(t + 1) + x_i(t - 1)] \sum_{i=1}^{n} a_{ij} x_j(t) = 0 \quad \text{iff} \quad x_i(t + 1) \ne x_i(t - 1)$$

and

$$-[x_i(t + 1) + x_i(t - 1)] \sum_{i=1}^{n} a_{ij} x_j(t) < 0 \quad \text{iff} \quad x_i(t + 1) = x_i(t - 1)$$

Furthermore, from the previous property it is easy to see that the network has only cycles of length 4 and, in a way similar to preceding sections, the transient, T, is bounded by:

$$T \le \|A\|/2$$

Example 7   For the antisymmetric matrix A defined by:

$$A = \begin{pmatrix} 0 & 1 & -1 & -1 \\ -1 & 0 & 1 & -1 \\ 1 & -1 & 0 & 1 \\ 1 & 1 & -1 & 0 \end{pmatrix}$$

we have the cycle:

$$-11-1-1 \to 11-11 \to -1-11-1 \to 1-111 \to -11-1-1 \to \ldots$$

4.8   Conclusion

The results presented here show how the dynamical behaviour of threshold automata networks closely resembles some physical systems such as spin glasses. In fact, as in a physical system, the global dynamics is driven by a discrete version of Lyapunov operators. This allowed us to prove fairly general results, a situation which is rather unusual in this subject.

Finally, previous analysis can be generalised to a large class of automata networks, *positive networks* [10]. For such networks there exist Lyapunov functions driving the dynamics.

4.9 **References**

[1] E. Bienenstock, F. Fogelman Soulié and G. Weisbuch (eds), *Disordered Systems and Biological Organization*, NATO ASI Series, Computer and System Science, F20, Springer Verlag, Berlin, Heidelberg, New York (1986)

[2] E. R. Berlekamp, J. H. Conway and R. K. Guy, *Winnings Ways for your Mathematical Plays*, vol. 2, chap. 25, Academic Press, New York (1982)

[3] J. Demongeot, E. Goles and M. Tchuente, *Dynamical Systems and Cellular Automata. Proc. Workshop Luminy*, Academic Press, New York (1985)

[4] F. Fogelman Soulié, Contribution à une théorie du calcul sur réseaux, Thesis, Grenoble (1985)

[5] F. Fogelman Soulié, E. Goles and G. Weisbuch, Transient length in sequential iteration of threshold functions, *Disc. Applied Maths.*, 6 (1983), 95–8

[6] F. Fogelman Soulié and E. Goles Chacc, Knowledge representation by automata networks, in *Computers and Computing*, P. Chenin, C. Di Crescenzo and F. Robert (eds), Coll. études et recherches en informatique, Masson (1986), 175–80

[7] E. Goles, and J. Olivos, The convergence of symmetric threshold automata, *Inf. & Control*, 5 (1981)

[8] E. Goles, and G. Vichniac, Lyapunov functions for parallel neural networks, Research Report, MIT Lab. for Computer Science, also in *Neural Networks for Computing*, J. S. Denker (ed.) *American Inst. of Physics*, no. 151 (1986)

[9] E. Goles, Comportement Dynamique de réseaux d'Automates, Thesis, Grenoble (1985)

[10] E. Goles, Dynamics on positive automata networks, *Theor. Comp. Sci.* 41 (1985)

[11] J. J. Hopfield, Neural networks and physical systems with emergent collective computational abilities, *Proc. Nat. Acad. Sci. USA*, 79 (1982), 2554–8

[12] D. Marr and T. Poggio, Cooperative computation of stereo disparity, *Science*, 194 (1976) 283–7

[13] W. McCulloch and W. Pitts, A logical calculus of the ideas immanent in nervous activity, *Bull. Math. Biophysics*, 5 (1943), 115–33

[14] J. von Neumann, *Theory of Self-reproducing Automata*, W. Burks (ed.), University of Illinois Press, Urbana (1966)

[15] S. Poljak and D. Turzik, On pre-period of discrete influence systems, Res. Report, Dept. of Systems Ing., U. Czechoslovakia (1984)

[16] J. B. Pollack and D. L. Waltz, Interpretation of natural language, *Byte*, (Feb. 1986), 189–98

[17] M. Tchuente, Contribution à l'étude des méthodes de calcul pour des systèmes de type coopératif, Thesis, Université de Grenoble, France (1982)

5 *Michel Cosnard and Driss Moumida*

Dynamical properties of an automaton with memory

5.1 Introduction

At the end of Chapter 2, the dynamics of an automaton with an unbounded memory was studied, by showing the equivalence between this automaton and the iteration of a continuous function of the circle. In this chapter, we study the dynamical behaviour of the same automaton but with the restriction of a bounded memory. The equation of the automaton is the following:

$$x_{n+1} = \mathbf{1}\left[t + \sum_{i=0}^{k-1} a_i x_{n-i} \right] \tag{5.1}$$

where

(i) x_n is the state of the automaton at the discrete time step n; it is a boolean variable;

(ii) a_i are the coupling coefficients; they are real constants;

(iii) t is the threshold; it is a real parameter;

(iv) k is the size of the memory; it is an integer constant;

(v) $\mathbf{1}[u] = 0$ if $u < 0$ and $\mathbf{1}[u] = 1$ if $u \geq 0$.

This automaton has been essentially introduced in order to model elementary properties of the nervous system [2, 3, 4, 18, 19]. Equation (5.1) is called a single neuronic equation and models the behaviour of a single neuron. In this case, the resting state is represented by 0, and the excited state by 1.

The general dynamic of such a model is extremely rich. Indeed it is shown in [18] and [19] that if we connect various automata of this kind – called *formalised neurons* – we can simulate every finite automaton. Thus such nets can carry out the control operation of a Turing machine.

This model has been studied by many authors for various choices of the coupling coefficients a_i and the threshold t. In this chapter, we present some results concerning these particular choices and we give some conjectures concerning the general case. More precisely, we study the cases in which the coefficients a_i form a geometric sequence and a palindromic word. Then we characterise the reversible automata according to the values of a_i and t.

Clearly Equation (5.1) is completely defined by the pair $(\mathbf{a}, t)$ where $\mathbf{a} = (a_0, a_1, \ldots, a_{k-1})$. It can be transformed into a system of order k in the classical way:

$$T: \{0, 1\}^k \rightarrow \{0, 1\}^k \tag{5.1'}$$

$$y(n) = (y_1(n), y_2(n), \ldots, y_k(n))$$
$$\mapsto y(n + 1) = T(y(n)) = (y_2(n), y_3(n), \ldots, y_k(n), f(y(n)))$$

where

$$f(y(n)) = \mathbf{1}\left[t + \sum_{i=0}^{k-1} a_i y_{k-i}(n) \right]$$

Hence, the automaton is equivalent to a discrete iteration on $\{0, 1\}^k$. Using this equivalence, (5.1) can also be seen to be an automata network. Indeed, let $T_1, \ldots, T_k$ be the components of T: $y_i(n + 1) = T_i(y(n))$. Clearly,

$$y_k(n + 1) = T_k(y(n)) = f(y(n)),$$

and for $1 \leq i \leq k - 1$,

$$y_i(n + 1) = T_i(y(n)) = y_{i+1}(n) = \mathbf{1}[y_{i+1}(n) - \tfrac{1}{2}].$$

The network is composed of k threshold automata (without memory). Henceforth, the study of (5.1) can be included in the general framework of the finite networks of threshold automata. In the chapter on the dynamics of such automata, a general theory for these networks is presented in the case of a symmetric connection graph, namely:

$$y_i(n + 1) = \mathbf{1}\left[t_i + \sum_{j=1}^{k} m_{ij} y_j(n) \right]$$

where $A = (m_{ij})_{1 \leq i, j \leq k}$ is a real symmetric matrix; i.e., $m_{ij} = m_{ji}$, and $t_i \in \mathbb{R}$.

In [13] Goles gave a characterisation of the dynamic of such a network when the matrix A is real symmetric:

The period of such a system is less than or equal to 2.

On the other hand, it is easily seen that the previous system (5.1') can be written as follows:

$$y_i(n + 1) = \mathbf{1}\left[t_i + \sum_{j=1}^{k} m_{ij} y_j(n) \right]$$

where $A = (m_{ij})_{1 \leq i, j \leq k}$ is a real matrix defined by:

$$\begin{aligned}
m_{ij} &= 1 && \text{if } j = i + 1 \text{ and } i = 1, 2, \ldots, k - 1, \\
m_{ki} &= a_{k-i} && \text{for } i = 1, 2, \ldots, k, \\
m_{ij} &= 0 && \text{if } j \neq i + 1 \text{ or } i \neq k,
\end{aligned}$$

and $\mathbf{t} = (t_i)_{1 \le i \le k}$ with $t_i = -\frac{1}{2}$ for i in $\{1, 2, \ldots, k - 1\}$ and $t_k = t$.

Note that the matrix A is not symmetric. Very few results are known in this case. Moreover to each arc $(i + 1, i)_{i=1,\ldots,k}$ we associate a weight equal to 1, and to each arc $(i, k)_{i=1,\ldots,k}$ we associate a weight equal to a_{k-i}. See Fig. 5.1.

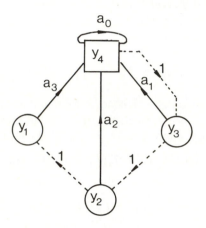

Figure 5.1 Directed graph associated to a single neuronic equation with memory of length $k = 4$ and coupling coefficient vector $\mathbf{a} = (a_0, a_1, a_2, a_3)$.

◯ The state of the cell i is the previous state of its neighbour cell $i + 1$.

▢ The state of the cell 4 is computed by taking into account all the cells of its neighbourhood: cells 1, 2, 3 and 4; with a threshold rule t.

5.2 Geometric memory

In this section we assume that the coupling coefficients form a geometric sequence: $a_i = -(b^i)$ with $b \in \,]0, \frac{1}{2}]$. This case has been studied in detail in [8]. Our presentation will follow that paper.

It is clear that the dynamical properties of equation (5.1) depend heavily on the particular sums Σb^i and the threshold t. In this case, we can easily show that equation (5.1) is equivalent to equation (5.2):

$$x_{n+1} = \mathbf{1}\left[t - \sum_{i=0}^{k-1} 2^i x_{n-i} \right] \tag{5.2}$$

The automaton of equation (5.2) will be denoted by $A(k, t)$. Here we recall some results whose complete proofs can be found in [8].

Theorem 1
(i) For every (k, t), $A(k, t)$ admits a unique globally attracting cycle. The period of this cycle is less than or equal to $k + 1$.
(ii) For every k, if $p \leq k + 1$, then there exists t such that $A(k, t)$ admits a p-cycle.

Sketch of the proof The proof of this theorem is very long. The structure of the cycles is characterised through a detailed analysis. In order to describe this structure we introduce the following substitution operator $S_{r,*}$ where r is an integer and $*$ is 0 or 1:

$$S_{r,0}: \begin{array}{c} 10^r \to 0 \\ 10^{r+1} \to 1 \end{array} \quad \text{and} \quad S_{r,1}: \begin{array}{c} 10^r \to 1 \\ 10^{r+1} \to 0 \end{array}$$

A given cycle C of $A(k, t)$ can be considered as a word on $\{0, 1\}$. Conversely, let C be such a word. There exists (k, t) such that C is a cycle of $A(k, t)$ if and only if we can define a sequence $(r_1, u_1), \ldots, (r_n, u_n)$ such that:

$$S_{r_n, u_n} \circ S_{r_{n-1}, u_{n-1}} \circ \ldots \circ S_{r_1, u_1}(C) = 0$$

The result follows from this characterisation. Note that this proof provides us with an algorithm for constructing the cycles of a given automaton. □

Let k be a fixed integer. For a given t there exists a unique cycle $C(k, t)$ for $A(k, t)$. We can describe the bifurcation structure of $A(k, t)$ completely as t varies. For this we associate to each cycle a rotation number defined by

$$\rho(k, t) = \frac{\text{number of 1 in } C(k, t)}{\text{length of } C(k, t)}$$

The unicity of the cycle for a given (k, t) implies that $\rho(k, t)$ is uniquely defined. Henceforth we will denote the rotation number either by $\rho(C)$ or $\rho(k, t)$.

Example 1 Let $k = 4$, $t = 0.5$ and $A(k, t)$ be defined as:

$$x_{n+1} = \mathbf{1}[0.5 - (x_n + 0.5x_{n-1} + 0.25x_{n-2} + 0.125x_{n-3})]$$

Consider the associated transformation $T: \{0, 1\}^k \to \{0, 1\}^k$

$$x = (x_0, x_1, \ldots, x_{k-1}) \mapsto z = T(x) = (x_1, \ldots, x_{k-1}, f(x))$$

and associate to each element $(x_0, x_1, \ldots, x_{k-1})$ of $\{0, 1\}^k$ the integer whose binary expansion is $(x_0, x_1, \ldots, x_{k-1})$. Thus T can be considered as a selfmap of $\{0, 1, \ldots, 2^k - 1\}$. The iteration graph of this automaton is depicted in Fig. 5.2. We can see that $A(k, t)$ admits a cycle of length 5, namely 00101. The rotation number is an irreducible fraction $\frac{2}{5}$.

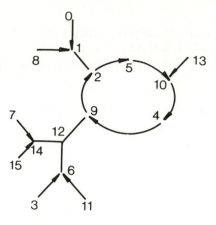

Figure 5.2 Iteration graph of example 1

Theorem 2

$\rho(k, .)$ is a monotone increasing surjection from $\mathbb{R}$ onto the irreducible fractions with denominators less than $k + 2$.

Sketch of the proof If C is the cycle of the automaton $A(k, t)$ then the rotation number $\rho(C)$ is an irreducible fraction. Conversely let p/q be a given irreducible fraction whose denominator is not greater than $k + 1$. Then to p/q we can associate a cycle C. The construction of such a cycle is given by the substitution defined in Theorem 1.   ☐

Table 5.1 and Fig. 5.2 illustrate Theorem 2.

t_{min}	t_{max}	C	T	$p/q = \rho(C)$
0	t^6	10^7	8	1/8
t^6	t^5	10^6	7	1/7
t^5	t^4	10^5	6	1/6
t^4	t^3	10^4	5	1/5
t^3	$t^2 + t^6$	10^3	4	1/4
$t^2 + t^6$	$t^2 + t^5$	$10^3 10^2$	7	2/7
$t^2 + t^5$	$t + t^4$	10^2	3	1/3
$t + t^4$	$t + t^4 + t^6$	$10^2 10^2 10$	8	3/8
$t + t^4 + t^6$	$t + t^3 + t^6$	$10^2 10$	5	2/5
$t + t^3 + t^6$	$t + t^3 + t^5$	$10^2 10\ 10$	7	3/7
$t + t^3 + t^5$	$t + t^2 + t^4 + t^6$	10	2	1/2

Table 5.1 $\rho(k, .)$ is constant on each interval $[t_{min}, t_{max}]$.

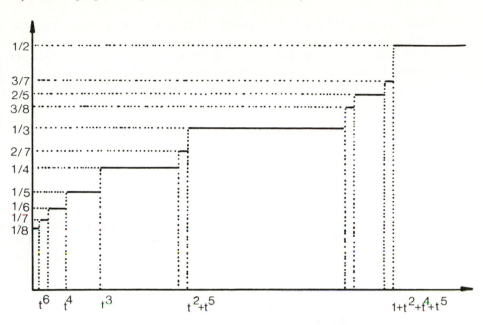

Figure 5.3 Graph of the function $\rho(7, .)$

The preceding results give us a way to obtain the bifurcation structure of equation (5.2). Consider the set of irreducible fractions with denominators not greater than $k + 1$ and list them in increasing order – we obtain a Farey sequence. The list of their denominators is the ordered list of the periods of the attractive cycles of (5.2). Using their numerators we can compute the values of t_{max}. Hence given a value of t, we can easily deduce the period of the corresponding cycle.

From a more general point of view, the theorems imply that a simple iteration formula can lead to a very rich structure of bifurcations. Note, however, that the length of the cycles is linear in k compared to the 2^k possible states.

5.3 Palindromic memory

We assume in this section that the coupling coefficients of (5.1) form a palindromic word, that is $a_i = a_{k-1-i}$ for every $i \in \{0, 1, \ldots, k - 1\}$. Such an automaton will be denoted $A(\mathbf{a}, t)$.

For every pair $(\mathbf{a}, t)$ with $\mathbf{a} = (a_0, a_1, \ldots, a_{k-1})$ there exists a real number ε such that the two systems $A(\mathbf{a}, t)$ and $A(\mathbf{a}, t + \varepsilon)$ are equivalent. Moreover $A(\mathbf{a}, t + \varepsilon)$ is a strict threshold automaton, that is,

for every $\mathbf{x} = (x_0, x_1, \ldots, x_{k-1}) \in \{0, 1\}^k$, we have $\sum_{i=0}^{k-1} a_i x_i + t + \varepsilon \neq 0$

Hence in this section we shall assume that $A(\mathbf{a}, t)$ is a strict threshold automaton.

Following Fogelman, Goles and Weisbuch [9], we can introduce an operator E characterising the dynamical behaviour of such systems:

for each $x = (x_0, x_1, \ldots, x_k) \in \{0, 1\}^{k+1}$, we define the operator E by

$$E(\mathbf{x}) = \sum_{j=0}^{-k-1} x_{k-j} \sum_{s=j+1}^{k} a_{s-j-1} x_{k-s} - t \sum_{j=0}^{k} x_j$$

More generally for a trajectory $(x_i)_{i \in \mathbb{N}}$ we have:

$$E(x_{n-k}, x_{n-k+1}, \ldots, x_n) = \sum_{j=0}^{-k-1} x_{n-j} \sum_{s=j+1}^{k} a_{s-j-1} x_{n-s} - t \sum_{j=0}^{k} x_{n-j}$$

The operator E can be seen as a kind of energy of the system. Indeed E is the Lyapunov function (for a more general approach see Goles [13]). The variation of E gives us an idea of how the system reaches a stationary state. For this purpose, we define a quantity Δ_n by

$$\Delta_n = E(x_{n-k}, \ldots, x_n) - E(x_{n-k-1}, \ldots, x_{n-1})$$

Let $(x_i)_{i \in \mathbb{N}}$ be a trajectory of the system whose period and transient will be denoted by p and q respectively.

Since the coupling coefficients a_i present a palindromic structure we get:

$$\Delta_n = E(x_{n-k}, \ldots, x_n) - E(x_{n-k-1}, \ldots, x_{n-1})$$

$$= (x_{n-k-1} - x_n)\left(t + \sum_{s=0}^{k-1} a_s x_{n-s-1}\right)$$

Lemma 1 If $x_n \neq x_{n-k-1}$ then $\Delta_n < 0$.

Proof Suppose that $x_n \neq x_{n-k-1}$. If $x_n = 1$ (which implies that $x_{n-k-1} = 0$) then

$$t + \sum_{s=0}^{k-1} a_s x_{n-s-1} > 0$$

If $x_n = 0$, then the proof is the same.   □

Theorem 3
The period p of a cycle of a given palindromic memory system $A(\mathbf{a}, t)$ divides $k + 1$.

Proof Note that for every i we have $\Delta_i \leq 0$. Since $x_{q+i} = x_{q+p+i}$ for $i = 1, 2, \ldots, k$, we deduce that

$$\sum_{i=1}^{p} \Delta_{q+k+i} = E(x_q, x_{q+1}, \ldots, x_{q+k}) - E(x_{q+p}, x_{q+p+1}, \ldots, x_{q+k+p}) = 0$$

Hence $\Delta_{q+k+i} = 0$ for every $i = 1, 2, \ldots, p$. Lemma 1 implies that $x_{k+q+i} = x_{q+i-1}$ for every $i = 1, 2, \ldots, p$ which proves that p divides $k + 1$. $\square$

The preceding operator can be used in order to derive a general bound for the length of the transient in the case where the coupling coefficients are integers.

Theorem 4

Let $A(\mathbf{a}, t)$ be a given palindromic memory system whose coupling coefficients are integers. The transient q of a trajectory $(x_i)_{i \in \mathbb{N}}$ of $A(\mathbf{a}, t)$ is bounded by:

$$q \le (k + s)^2 \left[2|t| + \sum_{s=0}^{k-1} |a_s| \right]$$

Proof   It is easily seen that for every $\mathbf{y} = (y_0, y_1, \ldots, y_k) \in \{0, 1\}^{k+1}$ we have

$$-\sum_{\substack{j=0 \\ a_s > 0}}^{k-1} \sum_{s=j+1}^{k} a_{s-j-1} \le E(y) \le t(k + 1) - \sum_{\substack{j=0 \\ a_s < 0}}^{k-1} \sum_{s=j+1}^{k} a_{s-j-1} \quad \text{if } t \ge 0$$

and

$$-\sum_{\substack{j=0 \\ a_s > 0}}^{k-1} \sum_{s=j+1}^{k} a_{s-j-1} + t(k + 1) \le E(y) \le -\sum_{\substack{j=0 \\ a_s < 0}}^{k-1} \sum_{s=j+1}^{k} a_{s-j-1} \quad \text{if } t \le 0$$

By applying this formula respectively to the vectors $(x_q, x_{q+1}, \ldots, x_{q+k})$ and $(x_0, x_1, \ldots, x_k)$ we obtain

$$|E(x_k, x_{k+1}, \ldots, x_{k+q}) - E(x_0, x_1, \ldots, x_k)|$$

$$\le |t|(k + 1) + \sum_{j=0}^{k-1} \sum_{s=j+1}^{k} |a_{s-j-1}|$$

Applying Lemma 1, we deduce that:

$$E(x_q, x_{q+s}, \ldots, x_{q+k}) - E(x_0, x_s, \ldots, x_k) = \sum_{i=1}^{q} \Delta_{k+i} < 0$$

Let us put $\delta = \min\{(-\Delta_{k+i}) \text{ such that } \Delta_{k+i} < 0 \text{ and } i = 1, 2, \ldots, q\}$

On the other hand we have

$$\delta = \min_{\substack{i=1,\ldots,q \\ x_{i-1} \neq x_{k+i}}} \left(|x_{i-1} - x_{k+i}| \cdot \left| t + \sum_{s=0}^{k-1} a_s \cdot x_{k+i-s-1} \right| \right)$$

$$= \min_{\substack{i=1,\ldots,q \\ \Delta_{k+1} \neq 0}} \left(\left| t + \sum_{s=0}^{k-1} a_s x_{k+i-1} \right| \right) \quad \text{since } |x_{k+i} - x_i| = 1$$

$$\Rightarrow q \cdot \delta \le \left(|t|(k + 1) + \sum_{j=0}^{k-1} \sum_{s=j+1}^{k} |a_{s-j-1}| \right)(k + 1)$$

Hence we can write

$$q \le (1/\delta)\left(|t| \cdot (k + 1) + \sum_{j=0}^{k-1} \sum_{s=j+1}^{k} |a_{s-j-1}|\right)(k + 1)$$

Since the coefficients a_i are integers we can always assume that we have

$$\left| t + \sum_{s=0}^{k-1} a_s x_{k+i-1} \right| \ge \tfrac{1}{2} \quad \text{for every integer } i;$$

which leads to the result.   □

The preceding analysis can be extended to the following case:

$$x_{n+1} = 1\left[t + \sum_{i=r}^{k-1} a_i x_{n-i} \right] \tag{5.3}$$

where the coefficients a_i are such that $a_i = a_{k+r-i-1}$ for $i \in \{r, r + 1, \ldots, k - 1\}$.

Lemma 2   The automaton defined by (5.3) is equivalent to the following

$$x_{n+1} = 1\left[t + \sum_{i=0}^{k+r-1} a'_i x_{n-i} \right] \tag{5.4}$$

$$\text{with} \quad a'_i = \begin{cases} a_i & \text{if } i = r, r + 1, \ldots, k - 1 \\ 0 & \text{otherwise} \end{cases}$$

Corollary 1
If $(x_i)_{i \in \mathbb{N}}$ is a trajectory of the system (5.4) with a transient length q and a period p, then we have:
  (i) the period p always divides $(k + r + 1)$,
  (ii) the length of the transient q is bounded by

$$q \le (k + r + 1)^2 \left[2|t| + \sum_{s=r}^{k-1} |a_s| \right]$$

Proof   The proof follows directly from Theorems 3 and 4.   □

Example 2   Let $k = 3$, $t = 2.5$ and $\mathbf{a} = (2, 1, 2)$

$$x_{n+1} = 1[2.5 - (2x_n + x_{n-1} + 2x_{n-2})]$$

We can see from Fig. 5.4 that $A(\mathbf{a}, t)$ does not possess a unique cycle and that the period of each cycle divides $(k + 1)$.
     We now consider a particular case of the palindromic memory. We shall assume that the coupling coefficients are of the form $a_{2i} = -1$ and $a_{2i+1} = 0$.

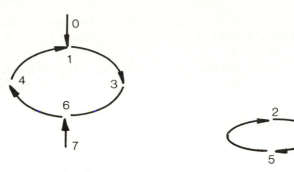

Figure 5.4 Iteration graph associated to example 2

Proposition 1

(i) If $t \geq [k/2]$ or $t < 0$ then $A(\mathbf{a}, t)$ has a unique globally attractive fixed point which is either 0 (if $t < 0$) or 1 (if $t \geq [k/2]$). $[z]$ is integral part of r.

(ii) If $0 \leqslant t < [k/2]$ then $A(\mathbf{a}, t)$ has a unique globally attractive cycle of order 2: $(0, 1)$.

Proof The first part of the proposition is obvious since the sign of $t + \Sigma\, a_i x_{n-i}$ is always the same.

In order to prove the second part, note that since $a_{2i+1} = 0$ we can assume that k is odd. Let $(x_0, x_1, \ldots, x_{k-1})$ be an initial configuration and let C be its associated cycle of period p. To prove that C is reduced to $(0, 1)$, we shall show that C does not have two consecutive 1 and that C does not have two consecutive 0.

Since k is odd we have:

$$x_{n+1} = \mathbf{1}[t - (x_n + x_{n-2} + x_{n-4} + \ldots + x_{n-k+1})] \tag{5.5}$$

Let us assume that C has two consecutive 1: there exists n such that $x_n = x_{n+1} = 1$ with $n \geq q$, where q denotes the length of the transient:

$$\begin{bmatrix} x_n = 1 \\ x_{n+1} = 1 \\ \vdots \\ x_{n+k-1} \end{bmatrix} \rightarrow \begin{bmatrix} x_{n+1} \\ x_{n+2} \\ \vdots \\ x_{n+k} \end{bmatrix} \rightarrow \begin{bmatrix} x_{n+2} \\ x_{n+3} \\ \vdots \\ x_{n+k+1} \end{bmatrix} \rightarrow \begin{bmatrix} x_{n+3} \\ x_{n+4} \\ \vdots \\ x_{n+k+2} \end{bmatrix}$$

Note that the vector $a = (a_0, a_1, a_2, \ldots, a_{k-1})$ is palindromic, which implies that p divides $(k + 1)$. Thus $x_n = x_{n+k+1} = 1$ and $x_{n+1} = x_{n+k+2} = 1$

On the other hand we have:

$$\begin{aligned} x_{n+k+1} &= \mathbf{1}[t - (x_{n+k} + x_{n+k-2} + \ldots + x_{n+1})] \\ &= \mathbf{1}[t - (x_{n+k+2} + x_{n+k} + x_{n+k-2} + \ldots + x_{n+3})] \\ &= x_{n+k+3} \end{aligned}$$

which proves that $x_{n+2} = x_{n+k+3} = 1$. With the same reasoning, we get $x_{n+k+r} = 1$. Hence $C = (1, 1, \ldots, 1)$ which gives a contradiction to the hypothesis $t < [k/2]$.

If we assume that C has two consecutive 0, we can easily show, in a similar way, that $x_{n+k+r} = 0$ for $r \geq 0$, which leads to the same conclusion.   □

5.4   Positive coupling coefficients

Definition 1

Let $\mathbf{x} = (x_1, x_2, \ldots, x_k)$ and $\mathbf{y} = (y_1, y_2, \ldots, y_k)$ be two vectors in $\{0, 1\}^k$

(i) We say that $\mathbf{x} \leq \mathbf{y}$ if $x_i \leq y_i$ for every i in $\{1, 2, 3, \ldots, k\}$;
(ii) We say that $\mathbf{x} < \mathbf{y}$ if $\mathbf{x} \leq \mathbf{y}$ and $\mathbf{x} \neq \mathbf{y}$;
(iii) $\mathbf{x}$ and $\mathbf{y}$ are comparable if $\mathbf{x} \leq \mathbf{y}$ or $\mathbf{y} \leq \mathbf{x}$.

If the coupling coefficients a_i are positive then the operator T defined as follows is a *monotone map:*

$$T: \mathbf{x}_n = (x_{n-k+1}, x_{n-k+2}, \ldots, x_n) \rightarrow (x_{n-k+2}, x_{n-k+3}, \ldots, x_n, f(\mathbf{x}_n))$$

where

$$f(\mathbf{x}_n) = \mathbf{1}\left[t + \sum_{i=0}^{k-1} a_i x_{n-i} \right]$$

Hence it is well known that if T generates a p-cycle $\mathbf{x}, T(\mathbf{x}), T^2(\mathbf{x}), \ldots, T^{p-1}(\mathbf{x})$ (with $p > 1$) then $\mathbf{x}, T(\mathbf{x}), T^2(\mathbf{x}), \ldots, T^{p-1}(\mathbf{x})$ are uncomparable.

Let $\mathbf{x}_n$ and $\mathbf{x}_m$ be two elements of a trajectory $(x_i)_{i \in \mathbb{N}}$, if $\mathbf{x}_n$ and $\mathbf{x}_m$ are comparable then we have a cycle whose length divides $|m - n|$. In particular if there exists an integer n such that $\mathbf{x}_n$ and $\mathbf{x}_{n+1}$ are comparable then the sequence $(x_i)_{i \in \mathbb{N}}$ converges towards a fixed point.

Proposition 2

If there exists an integer i such that $a_i + t \geq 0$ then the period of each cycle divides $i + 1$.

Proof   Note that if $x_{n-i} = 1$ then $x_{n+1} = 1$. Such an integer i will be called 1-pivot. Such systems will be studied, in detail, in Section 5.5.

Now let us see the particular case when the coupling coefficients are decreasing, that is,

$$0 \leq a_{k-1} \leq a_{k-2} \leq \ldots \leq a_{j+1} < a_j \leq \ldots \leq a_0$$

Proposition 3

If the coupling coefficients are decreasing positive then the automaton has only fixed points.

Proof We shall prove that if $x_n = 0$ then $x_{n+1} = 0$.

$$x_n = 0 \Leftrightarrow a_0 x_{n-1} + a_1 x_{n-2} + \ldots + a_{k-1} x_{n-k} + t < 0$$

$$a_1 x_{n-1} + a_2 x_{n-2} + \ldots + a_{k-1} x_{n-k+1} + a_k x_{n-k} + t$$
$$< a_0 x_{n-1} + a_1 x_{n-2} + \ldots + a_{k-1} x_{n-k} + t < 0$$

$$\Rightarrow a_0 x_n + a_1 x_{n-1} + \ldots + a_{k-1} x_{n-k+1} + t < 0 \qquad \square$$

If the coupling coefficients are increasing positive then we do not have the same result as before.

Example 3

$$x_{n+1} = \mathbf{1}[\dot{x}_n + 2x_{n-1} + 4x_{n-2} + 6x_{n-3} - 5.5] \qquad (5.6)$$

As shown in fig. 5.5, equation (5.6) admits two cycles of period 4 and 2 and two fixed points. Note that the period of each cycle divides k. This is a direct consequence of Proposition 2 since we have $a_{k-1} + t \le 0$.

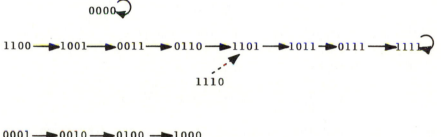

Figure 5.5 Iteration graph associated to example 3

5.5 Reversible systems [23]

Let T be the operator defined on $\{0, 1\}^k$ by

$$x = (x_0, x_1, \ldots, x_{k-1}) \rightarrow T(x) = (x_1, x_2, \ldots, x_{k-1}, f(x))$$

associated to the automaton $A(\mathbf{a}, t)$.

Definition 2

$A(\mathbf{a}, t)$ is a *reversible automaton* if T is a bijection.

Definition 3

$A(\mathbf{a}, t)$ is a *shift* if $f(x_0, x_1, \ldots, x_{k-1}) = x_0$, that is:

$$T(x_0, x_1, \ldots, x_{k-1}) = (x_1, x_2, \ldots, x_{k-1}, x_0)$$

$A(\mathbf{a}, t)$ is an *antishift* if $f(x_0, x_1, \ldots, x_{k-1}) = 1 - x_0$.

Clearly, if $A(\mathbf{a}, t)$ is a shift then for every $\mathbf{x}$ in $\{0, 1\}^k$ we have $T^k(\mathbf{x}) = \mathbf{x}$, and if $A(\mathbf{a}, t)$ is an antishift then we have $T^{2k}(\mathbf{x}) = \mathbf{x}$. We deduce from this remark that if $A(\mathbf{a}, t)$ is a shift or an antishift then $A(\mathbf{a}, t)$ is reversible. Moreover we have $T^{-1} = T^{k-1}$ if $A(\mathbf{a}, t)$ is a shift and $T^{-1} = T^{2k-1}$ if $A(\mathbf{a}, t)$ is an antishift.

Proposition 4

$A(\mathbf{a}, t)$ is reversible if and only if $A(\mathbf{a}, t)$ is a shift or an antishift.

Proof We shall prove that if $A(\mathbf{a}, t)$ is reversible then $A(\mathbf{a}, t)$ is a shift or an antishift. We distinguish three cases: $a_{k-1} > 0$, $a_{k-1} < 0$, $a_{k-1} = 0$.

If $a_{k-1} > 0$ then we have

$$t + \sum_{i=0}^{k-2} a_i x_{k-1-i} < t + \sum_{i=0}^{k-2} a_i x_{k-1-i} + a_{k-1}$$

which implies that $f(0, x_1, x_2, \ldots, x_{k-1}) = 0$ and $f(1, x_1, x_2, \ldots, x_{k-1}) = 1$. Hence $A(\mathbf{a}, t)$ is a shift.

If $a_{k-1} < 0$, we have:

$$t + \sum_{i=0}^{k-1} a_i x_{k-1-i} + a_{k-1} < t + \sum_{i=0}^{k-1} a_i x_{k-1-i}$$

which implies that $f(0, x_1, x_2, \ldots, x_{k-1}) = 1$ and $f(1, x_1, x_2, \ldots, x_{k-1}) = 0$. Hence, $A(\mathbf{a}, t)$ is an antishift.

If $a_{k-1} = 0$ then T cannot be a bijection. Indeed for two different elements, we have:

$$f(0, x_1, x_2, \ldots, x_{k-1}) = f(1, x_1, x_2, \ldots, x_{k-1})$$

Thus T is not a bijection. $\qquad\square$

Proposition 5

(i) $A(\mathbf{a}, t)$ is a shift if and only if

$$\sum_{i \le k-2 \text{ and } a_i > 0} a_i + t < 0 \quad \text{and} \quad \sum_{i \le k-2 \text{ and } a_i < 0} a_i + a_{k-1} + t \ge 0 \quad (5.7)$$

(ii) $A(\mathbf{a}, t)$ is an antishift if and only if

$$\sum_{i \le k-2 \text{ and } a_i > 0} a_i + t \le 0 \quad \text{and} \quad \sum_{i \le k-2 \text{ and } a_i < 0} a_i + a_{k-1} + t < 0 \quad (5.8)$$

Proof Assume that $A(\mathbf{a}, t)$ is a shift. For $\mathbf{x} = (x_1, x_2, \ldots, x_{k-1})$ in $\{0, 1\}^{k-1}$

define $(\mathbf{x}, 0) = (0, x_1, x_2, \ldots, x_{k-1})$ and $(\mathbf{x}, 1) = (1, x_1, x_2, \ldots, x_{k-s})$. We have $f(\mathbf{x}, 0) = 0$ and $f(\mathbf{x}, 1) = 1$.

Define $\quad \mathscr{P} = \{i \in \{0, 1, \ldots, k - 2\}$ such that $a_i > 0\}$
$\quad\quad\quad\; \mathscr{N} = \{i \in \{0, 1, \ldots, k - 2\}$ such that $a_i < 0\}$

Let $\mathbf{y} = (y_1, y_2, \ldots, y_{k-1}) \in \{0, 1\}^{k-1}$ be such that

$$y_{k-1-i} = \begin{cases} 0 & \text{if } i \in \mathscr{N} \\ 1 & \text{if } i \in \mathscr{P} \\ * & \text{otherwise} \end{cases}$$

where $y_i = *$ means that y_i can take any value in $\{0, 1\}$,

$$f(\mathbf{y}, 0) = 1\left[\sum_{i \le k-2 \text{ and } a_i > 0} a_i + t\right] \text{ implies that } \sum_{i \le k-2 \text{ and } a_i > 0} a_i + t < 0$$

Let now $\mathbf{z} = (z_1, z_2, \ldots, z_{k-1})$ be such that $z_i = 1 - y_i$.

Since $f(\mathbf{z}, 1) = 1$ we obtain $1\left[a_{k-1} + \sum_{i \le k-2 \text{ and } a_i < 0} a_i + t\right] = 1$ which implies that

$$\sum_{i \le k-2 \text{ and } a_i < 0} a_i + a_{k-1} + t \ge 0$$

Conversely, let us assume that the formula (5.7) is verified. We shall prove that $A(\mathbf{a}, t)$ is a shift.

Since $\sum_{i=0}^{k-2} a_i x_{k-1-i} + t \le \sum_{i \le k-2 \text{ and } a_i > 0} a_i + t < 0$, we deduce that

$f(\mathbf{x}, 0) = 0$.

In a similar way we can show that $f(\mathbf{x}, 1) = 1$ $\qquad\qquad\square$

Theorem 5
(i) A reversible automaton $A(\mathbf{a}, t)$ has only cycles of length L such that:
  (a) L divides k if $A(\mathbf{a}, t)$ is a shift (i.e. $a_{k-1} > 0$),
  (b) L divides $2k$ if $A(\mathbf{a}, t)$ is an antishift (i.e. $a_{k-1} < 0$),
(ii) The automaton defined by equation (5.1) cannot have a cycle of length 2^k.

Proof

(i) This is an immediate consequence of the previous remarks.
(ii) If $A(\mathbf{a}, t)$ has a cycle of length $L = 2^k$, then the associated operator T also has a cycle of length $L = 2^k$, and thus is bijective leading to a contradiction when k is greater than 1. $\qquad\qquad\square$

Many properties of shift and antishift operators are given in [17]. Moreover note that if we have an automaton with a geometric memory such that $b \leq 1$ then the operator T is never bijective. But if we have $b > 1$ then there exists $t_{(b)}$ such that the operator T is bijective.

From Proposition 5 we deduce that there exist only two different reversible memory systems: the shift and the antishift. Hence if a_{k-1} is positive the reversible system is equivalent to $A(\mathbf{a}, t)$ with $\mathbf{a} = (0, 0, 0, \ldots, 0, 1)$ and $t = -\frac{1}{2}$ (shift). Otherwise if a_{k-1} is negative the reversible system is equivalent to $A(\mathbf{a}, t)$ with $\mathbf{a} = (0, 0, 0, \ldots, 0, -1)$ and $t = \frac{1}{2}$ (antishift).

5.6 Pivot sum

We shall first study the case of a single pivot and then attempt to discuss the general case.

Definition 4

i is *0-pivot* if for every n, $x_{n-i} = 0$ implies $x_{n+1} = 0$.
i is *1-pivot* if for every n, $x_{n-i} = 1$ implies $x_{n+1} = 1$.
i is an *antipivot* if for every n, $x_{n-i} = 0$ is equivalent to $x_{n+1} = 1$.

Lemma 3

i is 0-pivot if and only if

$$\sum_{\substack{j=0 \\ j \neq i \text{ and } a_j > 0}}^{k-1} a_j + t < 0$$

i is 1-pivot if and only if

$$a_i + \sum_{\substack{j=0 \\ j \neq i \text{ and } a_j < 0}}^{k-1} a_j + t \geq 0$$

i is an antipivot if and only if

$$a_i + \sum_{\substack{j=0 \\ j \neq i \text{ and } a_j > 0}}^{k-1} a_j < -t \leq \sum_{\substack{j=0 \\ j \neq i \text{ and } a_j < 0}}^{k-1} a_j$$

Proof Follows from direct computations.

Theorem 6

Let p be the period of a cycle of the automaton $A(\mathbf{a}, t)$.
If i is a 0-pivot or a 1-pivot then p divides $(i + 1)$.
If i is an antipivot then p divides $2(i + 1)$.

Proof Assume that i is a 0-pivot. In order to prove that p divides $(i + 1)$, we introduce the following operator:

$$E_i(n) = x_{n-i} - x_{n+1}$$

It is clear that E_i is a positive function, that is, for every n, $E_i(n) \geq 0$. Denote by q the length of the transient of the trajectory and assume that there exists an integer $n > q + i$ such that $E_i(n) > 0$. We then have $x_{n+1} < x_{n-i}$. Since E_i is a positive function, we have:

$$x_{n-i+p(i+1)} \leq x_{n-i+(p-1)(i+1)} \leq \cdots \leq x_{n+1} < x_{n-i}$$

Since p is the period of the trajectory we have $x_{n-i+p(i+1)} = x_{n-i}$. Thus we get a contradiction. Hence E_i can be seen as an invariant operator on each cycle of the system.

The proof for the 1-pivot and the antipivot is similar. $\qquad\square$

We shall now treat two cases of pivot sum:

Definition 5
Let $\mathcal{T}$ be a subset of $\{0, 1, \ldots, k - 1\}$.
(i) $\mathcal{T}$ represents 0-pivot sum if

$$x_{n+1} = 1 \Rightarrow \forall i \in \mathcal{T} \quad x_{n-i} = 1$$

(ii) $\mathcal{T}$ represents 1-pivot sum if

$$x_{n+1} = 0 \Rightarrow \forall i \in \mathcal{T} \quad x_{n-i} = 0$$

Lemma 4
(i) $\mathcal{T}$ represents 0-pivot sum if and only if

$$\sup_{i \in \mathcal{T}} \left[\sum_{\substack{j=0 \\ j \neq i \text{ and } a_j > 0}}^{k-1} a_j + t \right] < 0$$

(ii) $\mathcal{T}$ represents 1-pivot sum if and only if

$$\inf_{i \in \mathcal{T}} \left[a_i + \sum_{\substack{j=0 \\ j \neq i \text{ and } a_j < 0}}^{k-1} a_j + t \right] \geq 0$$

Proof Follows from a direct computation.

Corollary 2
If we have 0-pivot sum or 1-pivot sum then the period p divides $\gcd_{i \in \mathcal{T}}(i + 1)$, where $\gcd$ = greatest common divisor.

Proof Follows directly from the fact that each $i \in \mathcal{T}$ constitutes a pivot of the system.

Corollary 3

Let $\mathcal{T}$ represent τ-pivot sum (with $\tau = 0$ or $\tau = 1$).
If $\mathcal{T}$ contains a prime integer predecessor or two consecutive integers then the system has only fixed points.

Proof Obvious.

Remarks

Recently Tchuente [25] has given another way to see the neuronic equation (5.1). For a given weight vector $\mathbf{a} = (a_0, a_1, \ldots, a_{k-1})$ he associates a real matrix $\mathbf{A} = (A_{ij}) \in M_{(k+1)(k+1)}(\mathbb{R})$ defined by

$$A_{ij} = \begin{cases} a_{j-i-1} & \text{if } i < j \\ 0 & \text{if } i = j \\ a_{k+j-i} & \text{if } i > j \end{cases}$$

Note that $\mathbf{A}$ is symmetric if and only if $\mathbf{a} = (a_0, a_1, \ldots, a_{k-1})$ is a palindromic vector.

Now we will give some definitions and notations.
Let φ be a boolean function defined on $\{0, 1\}^{k+1}$ by

$$\varphi: \{0, 1\}^{k+1} \rightarrow \{0, 1\}^{k+1}$$

$$\mathbf{s} = (s_1, s_2, \ldots, s_{k+1}) \mapsto \varphi(\mathbf{s}) = (\varphi_1(\mathbf{s}), \varphi_2(\mathbf{s}), \ldots, \varphi_{k+1}(\mathbf{s}))$$

where

$$\varphi_i(\mathbf{s}) = 1\left[\sum_{j=0}^{k+1} A_{ij}s_j + t\right]$$

where t is the threshold of the neuronic equation (5.1).

Let $y_i(n)$ denote the state of the cell i at the time step n, and $y(n) = (y_1(n), y_2(n), \ldots, y_{k+1}(n))$ be the configuration of the network at time step n. Note that we have $k + 1$ cells.

We define a sequence $(z_n)_{n \in \mathbb{N}}$ by

$z_{(k+1)n+i-1} = y_i(n)$ where $i = 1, 2, \ldots, k + 1$, and where $z_0 = *$ (z_0 takes any value in $\{0, 1\}$).

$y_i(n)$, $i = 1, 2, \ldots, k + 1$, are such that

$$y_1(n + 1) = \psi_1(y(n)) = \varphi_1(y_1(n), y_2(n), \ldots, y_{k+1}(n))$$
$$y_2(n + 1) = \psi_2(y(n)) = \varphi_2(y_1(n + 1), y_2(n), y_3(n), \ldots, y_{k+1}(n))$$
$$\vdots$$
$$y_i(n + 1) = \psi_i(y(n))$$
$$\qquad\qquad = \varphi_i(y_1(n + 1), y_2(n + 1), \ldots, y_{i-1}(n + 1), y_i(n), \ldots, y_{k+1}(n))$$
$$\vdots$$
$$y_{k+1}(n + 1) = \psi_{k+1}(y(n))$$
$$\qquad\qquad = \varphi_{k+1}(y_1(n + 1), y_2(n + 1), \ldots, y_k(n + 1), y_{k+1}(n))$$

Thus we can write $y(n + 1) = \psi(y(n))$ where $\psi = (\psi_1, \psi_2, \ldots, \psi_{k+1})$, defined as above, corresponds to the sequential mode iteration of the function φ.

Proposition 6

If **a** is a palindrome then for any initial configuration the neural network converges towards a fixed point.

Proof If **a** is a palindrome then the matrix **A** is real symmetric. Thus the result follows from a theorem of [13].

Let us return to the sequence $(z_i)_{i \in \mathbb{N}}$ defined above. Proposition 6 implies that there exists an integer q such that $y(n + 1) = y(n)$ for every integer $n \geq q$. Hence we have

$$z_{(k+1)n+i-1} = y_i(n) = y_i(n + 1) = z_{(k+1)(n+1)+i-1}.$$

Hence, we deduce that if the coupling coefficient vector **a** is palindromic then the period of the sequence $(z_i)_{i \in \mathbb{N}}$ always divides $k + 1$.

This constitutes another proof of Theorem 3.

5.7 Some conjectures

If the coupling coefficients a_i are positive then the period of each cycle of a given automaton $A(\mathbf{a}, t)$ is less than $k + 1$.

In the general case, the period of each cycle of a given automaton $A(\mathbf{a}, t)$ is less than $2k + 1$.

5.8 References

[1] J. P. Allouche and M. Cosnard, Itérations de fonctions unimodales et suites engendrées par automates, *C. R. Acad. Sci.*, 296 (1983), 159–62
[2] E. R. Caianiello, Decision equation and reverberations, *Kybernetik*, 3, 2 (1966)
[3] E. R. Caianiello and A. De Luca, Decision equation for binary systems. Application to neuronal behavior, *Kybernetik*, 3 (1966), 33–40
[4] E. R. Caianiello, A. De Luca and Ricciardi, Reverberation and control of neural networks, *Kybernetik*, 4, 1 (1967)
[5] P. Collet and J. P. Eckmann, *Iterated maps of the Interval as Dynamical Systems*, Birkhäuser, Basel, (1980)
[6] M. Cosnard and E. Goles, Comportement dynamique d'un automate à mémoire analogie avec le fonctionement d'un neurone, *IV° Séminaire de l'Ecole de Biologie Théorique* (1985)
[7] M. Cosnard and E. Goles, Dynamique d'un automate à mémoire modélisant le fonctionnement d'un neurone, *C. R. Acad. Sci.*, 299, 10 (1984), 459–61
[8] M. Cosnard and E. Goles, Bifurcation structure of discrete neuronal equation, Submitted to *Disc. Math.*
[9] F. Fogelman Soulié, E. Goles and G. Weisbush, Decreasing energy as a tool for studying threshold networks, *Disc. Appl. Math.*, 6 (1983), 95–8

[10] F. Fogelman Soulié, Contributions à une théorie du calcul sur réseaux, Thesis, Grenoble (1985)

[11] E. Goles, Fixed point behaviour of threshold functions on a finite set, *SIAM J. Alg. and Appl. Math.*, 3, 4, (Dec. 1982)

[12] E. Goles, Dynamical behaviour of neural networks, Res. Repp. 386, IMAG, Grenoble, France (1983)

[13] E. Goles Chacc, Comportement dynamique de réseaux d'automates, Thesis, Grenoble (1985)

[14] E. Goles and J. Olivos, Comportement périodique des fonctions à seuil binaires et applications, *Disc. Appl. Math.*, 3, (1984)

[15] E. Goles and M. Tchuente, Dynamic behaviour of one-dimensional threshold automata, Res. Rep. 350, IMAG, Grenoble, France (1983)

[16] T. Kitagawa, Cell space approach, *Math. Biosc.*, 19 (1974)

[17] T. Kitagawa, Dynamical systems and operators associated with a single neuronic equation, *Math. Biosc.*, 18 (1973)

[18] S. C. Kleene, Representation of events in nets and finite automata, *Automata Studies*, C. E. Shannon and J. MacCarthy (ed.), Princeton University Press, Princeton, NJ

[19] W. S. McCulloch and W. Pitts, A logical calculus of the ideas immanent in nervous activity, *Bull. Math. Biophysics*, 5 (1943), 115–33

[20] H. Nagami *et al.*, Characterization of dynamical behaviour associated with a single neural equation, *Math. Biosc.*, 18 (1973)

[21] J. Nagumo and S. Sato, On a response characteristic of a mathematical neuron model, *Kybernetik*, 3 (1972), 155–64

[22] Y. Robert, Algorithmique parallèle: réseaux d'automates, architectures systoliques, machines SIMD et MIMD, Thesis, Grenoble (1986)

[23] T. de Saint Pierre, Dinámica de autómatas, Thesis, Santiago, Chile (1985)

[24] M. Tchuente, Contribution à l'étude des méthodes de calcul pour des systèmes de type coopératif, Thesis, Grenoble (1984)

[25] M. Tchuente, Private communication

Computation on automata networks

6.1 Introduction

A deterministic *cellular automaton* is a model originally introduced by von Neumann [28], which consists of a collection of elementary automata with local interconnections, evolving in a parallel and synchronous way. For instance, a two-dimensional cellular automaton can be built up in the following way:

(i) We begin with a *cellular space* which consists of an infinite plane divided into squares indexed by pairs of integers, $(i, j) \in \mathbb{Z} \times \mathbb{Z}$.

(ii) There is a *neighbourhood* relation which gives for any cell $x = (i, j)$, a finite list of cells $x + v_1, x + v_2, \ldots, x + v_k$ called its neighbours and from which it can directly receive information; in the general case, the vector $V = (v_1, v_2, \ldots, v_k)$, which is usually called the neighbourhood index, may vary from one cell to another.

(iii) We specify a finite list of *states* Q, and a distinguished state called 'quiescent state' and denoted 0.

(iv) For any cell x of the cellular space, we define a *local transition function* f_x from Q^k to Q, where k denotes the number of neighbours of x; f_x specifies the state $q^{t+1}(x)$ of cell x at time $t + 1$, as a function of the states of its neighbours at time t, that is,

$$q^{t+1}(x) = f_x(q^t(x + v_1), \ldots q^t(x + v_k)).$$

(v) Any function $c: \mathbb{Z} \times \mathbb{Z} \to Q$ which defines an assignment of states to all cells in the cellular space is called a *configuration* and, for any cell x, $c(x)$ is called the state of x under configuration c. A configuration c is said to be *finite*, if all except a finite number of cells are quiescent.

(vi) The simultaneous application of the local transition functions f_x to all the cells of the cellular space, defines a *global transition function* F which acts on the entire array, transforming any configuration c into a new configuration c' such that

$$c'(x) = f_x(c(x + v_1), \ldots, c(x + v_k))$$

(vii) Usually, it is assumed that,

$$f_x(0, \ldots, 0) = 0, \text{ for any } x,$$

so that the image of any finite configuration is also finite.

6.1.1 *Infinite uniform cellular structures*

Definition   A cellular automaton is said to be *uniform* if the neighbourhood relation and the transition function are the same for every cell. In this case, the vector $V = (v_1, \ldots, v_n)$ is called the neighbourhood index, and the common local transition function is denoted by f.

This restriction prevents one from building a specific automaton for every cell of the cellular space. The model introduced by von Neumann was uniform, because one of his interests in introducing cellular automata was the development and organisation of biological systems under the adverse conditions of homogeneity and isotropy. The three examples presented below correspond to infinite uniform cellular structures.

Example 1

Let us consider a binary one-dimensional array with nearest-neighbour interactions, that is, a cellular space formally defined as follows (Fig. 6.1):

the set of cells is indexed by $\mathbb{Z}$
$Q = \{0, 1\}$
$V = (-1, 0, 1)$
for any cell of the array, the local transition function is the majority function.

$$f(a_{-1}, a_0, a_1) = \begin{cases} 1 \text{ if } a_{-1} + a_0 + a_1 \geq 2 \\ 0 \text{ otherwise} \end{cases}$$

Figure 6.1 Interconnection pattern

An example of evolution:

```
t = 0, c:       ... 0 1 0 1 0 1 0 1 1 0 0 1 0 1 1 0 0 ...
t = 1, F(c):    ... 0 0 1 0 1 0 1 1 1 0 0 0 1 1 1 0 0 ...
t = 2, F²(c)    ... 0 0 0 1 0 1 1 1 1 0 0 0 1 1 1 0 0 ...
t = 3, F³(c)    ... 0 0 0 0 1 1 1 1 1 0 0 0 1 1 1 0 0 ...
t = 4, F⁴(c)    ... 0 0 0 0 1 1 1 1 1 0 0 0 1 1 1 0 0 ...
```

The configuration at time $t = 3$ is stable, i.e. $F^3(c) = F^4(c)$.

Example 2

Let us consider a binary two-dimensional array defined as follows:

(i) The cells of the array are indexed by $\mathbb{Z} \times \mathbb{Z}$.

(ii) Any cell $x = (i, j)$ is directly connected to its four nearest neighbours i.e.
$V = [(0, 1), (0, -1), (1, 0), (-1, 0)]$

(iii) The local transition function is the same for all the cells of the array and is defined as follows:
$q^{t+1}(i, j) = q^t(i + 1, j) + q^t(i - 1, j) + q^t(i, j + 1) + q^t(i, j - 1)$
(mod 2).

Figure 6.2 Neighbourhood of the darkened cell (example 2)

```
...  0 0 0 0 0  ...   ...  0 0 0 0 0  ...   ...  0 0 1 0 0  ...

...  0 0 0 0 0  ...   ...  0 0 1 0 0  ...   ...  0 0 0 0 0  ...

...  0 0 1 0 0  ...   ...  0 1 0 1 0  ...   ...  1 0 0 0 1  ...

...  0 0 0 0 0  ...   ...  0 0 1 0 0  ...   ...  0 0 0 0 0  ...

...  0 0 0 0 0  ...   ...  0 0 0 0 0  ...   ...  0 0 1 0 0  ...
```

       $t = 0$ $t = 1$ $t = 2$

Figure 6.3 an example of evolution

Example 3

The game of Life [7].

(i) The cells are indexed by $\mathbb{Z} \times \mathbb{Z}$

(ii) $Q = \{0, 1\}$

(iii) $V = \{(i, j); -1 \le i \le 1, -1 \le j \le 1\}$ (see Fig. 6.4)

(iv) The local transition function is the same for all the cells of the array, and is defined as follows:

A cell in state 0 (resp. 1) is said to be dead (resp. alive).

Birth: a cell that is dead at time t becomes alive at time $t + 1$ only if exactly 3 of its neighbours were alive at time t

Death by overcrowding: a cell that is alive and has 4 or more neighbours alive at time t will be dead at time $t + 1$.

Death by exposure: a live cell that has only one live neighbour or none at all at time t, will also be dead at time $t + 1$.

Survival: a cell that was alive at time t will remain alive at time $t + 1$ if and only if it has just 2 or 3 neighbours alive at time t.

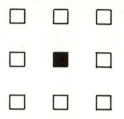

Figure 6.4 Neighbourhood of the darkened cell (example 3)

```
                                        1                    1
                      1 1 1           1 0 1              1 1 1
        1 1 1 1       1 1 1       1 0 0 0 1          1 1 0 1 1
                      1 1 1           1 0 1              1 1 1
                                        1                    1
          t = 0         t = 1           t = 2              t = 3


                                        1                  1 1 1
        0 1 1 1 0                 0 1 1 1 0        0 0 0 0 0 0 0
        1 0 0 0 1                 1 0 1 0 1        1 0 0 0 0 0 1
        1 0 0 0 1         1 1 1 0 1 1 1            1 0 0 0 0 0 1
        1 0 0 0 1                 1 0 1 0 1        1 0 0 0 0 0 1
        0 1 1 1 0                 0 1 1 1 0        0 0 0 0 0 0 0
                                        1                  1 1 1
          t = 4                    t = 5              t = 6


                      1
                      1
                      1                              1 1 1
        0 0 0 0 0 0 0 0 0                     0 0 0 0 0 0 0
        1 1 1 0 0 0 1 1 1                     1 0 0 0 0 0 1
        0 0 0 0 0 0 0 0 0                     1 0 0 0 0 0 1
                      1                       1 0 0 0 0 0 1
                      1                       0 0 0 0 0 0 0
                      1                         1 1 1
        t = 7: configuration c           t = 8: configuration c'
```

Figure 6.5 $F(c) = c'$ and $F(c') = c$ (traffic lights)

The second example of evolution (Fig. 6.6) shows a configuration called a 'Glider', which moves one square diagonally every four generations.

```
0 0 1 0 0

0 0 0 1 0   0 1 0 1 0   0 0 0 1 0   0 0 1 0 0   0 0 0 1 0

0 1 1 1 0   0 0 1 1 0   0 1 0 1 0   0 0 0 1 1   0 0 0 0 1

            0 0 1 0 0   0 0 1 1 0   0 0 1 1 0   0 0 1 1 1

   t = 0       t = 1       t = 2       t = 3       t = 4
```

Figure 6.6 A glider moving south-east

6.1.2 *Uniform nondeterministic cellular structures*

The preceding definitions concerned deterministic cellular automata, that is structures where for any given initial configuration, the transition function yields a unique next configuration and hence a unique history through time.

A *nondeterministic* cellular automaton starts with a given initial configuration, but its transition function yields a set of possible next states for each cell [29].

Let us define a pattern as a function P from $\mathbb{Z}^n$ into Q which is defined at finitely many points. The domain of a pattern is the set of points at which it is defined. We shall say that two patterns agree, or that a pattern agrees with a configuration, if they are equal on the intersection of their domains.

If P is a pattern and x is in $\mathbb{Z}^n$, then P^x denotes the pattern obtained by translating P by x, i.e. $P(y) = P^x(-x + y)$.

A finite neighbourhood condition, $P(x, c)$ is defined by a finite list of patterns $P_1, \ldots, P_k$ whose domains contain the origin O. For a configuration c and a cell x, we say that $P(x, c)$ is true if and only if c agrees with one of $P_1^x, \ldots, P_k^x$.

A uniform nondeterministic local transition function f consists of a finite list of admissible transitions $P_i(x, c) \mapsto r_i$ for $i = 1, \ldots, k$, where each $P_i(x, c)$ is a finite neighbourhood condition, and r_i is an element of Q.

The global transition function associated with T is a relation between configurations defined by:

$c'Tc''$ if and only if for any x in $\mathbb{Z}^n$, there is a production $P_i(x, c) \mapsto r_i$ such that $P_i(x, c')$ is true and $c''(x) = r_i$.

6.2 Construction and self-reproduction

As reported by Arbib [2], 'we intuitively associate with machines used for construction, a certain degenerating tendency: we expect an automaton to build an automaton of less complexity. However, when organisms reproduce, we expect their offspring to be of complexity at least equal to that of the parent. In fact, because of long-term processes of evolution, we even expect to see increases in complexity during reproduction. In view of this apparent conflict, von Neumann felt it worthwhile to see what could be formulated rigorously about the construction of one automaton by another.'

It is well known in theoretical computer science that, if an algorithm can be performed by any machine, then it can be realised by a Turing machine. However the concept of Turing machine is not powerful enough to model construction. Indeed, a Turing machine is only required to carry out logical manipulations on its tape, sense symbols, move its tape, print symbols and carry out logical elementary operations, whereas an automaton which constructs, must be able to recognise components, move them around and join them together. So von Neumann introduced the concept of cellular automata in which an automaton can be embedded by defining appropriately the states of a convenient finite subset of cells.

An automaton P is said to *construct* another automaton P' if, starting with a finite configuration c where all cells except those of P are in the 'quiescent state', there is a time t such that:

P' is a subconfiguration of $F'(c)$ and

$F'(c) - P'$ does not pass information to P'.

An example of a configuration that constructs is the 'glider gun' exhibited in 'Life' by R. W. Gosper and his research group at MIT in 1970, and which emits a new glider every 30 generations (see Fig. 6.7).

```
0 0 0 0 0 0 0 0 0 0 0 0 0 0 0 0 0 0 0 0 0 0 0 0 1
0 0 0 0 0 0 0 0 0 0 0 0 0 0 0 0 0 0 0 0 0 0 1 1 1 1
0 0 0 0 0 0 0 0 0 0 0 0 1 0 0 0 0 0 0 1 1 1 1 0 0 0 0 0 0 0 0 1 1
0 0 0 0 0 0 0 0 0 0 0 1 0 1 0 0 0 0 0 0 1 0 0 1 0 0 0 0 0 0 0 0 1 1
1 1 0 0 0 0 0 0 0 0 1 0 0 0 1 1 0 0 0 0 1 1 1 1 0 0 0 0 0 1
1 1 0 0 0 0 0 0 0 0 1 0 0 0 1 1 0 0 0 0 0 1 1 1 1 0 0 0 0 1
0 0 0 0 0 0 0 0 0 0 1 0 0 0 1 1 0 0 0 0 0 0 0 1
0 0 0 0 0 0 0 0 0 0 0 1 0 1
0 0 0 0 0 0 0 0 0 0 0 1
0 0 0 0 0 0 0 0 0 0 0 0 0 0 0 0 0 0 0 0 0 1 0 1
0 0 0 0 0 0 0 0 0 0 0 0 0 0 0 0 0 0 0 0 0 0 1 1
0 0 0 0 0 0 0 0 0 0 0 0 0 0 0 0 0 0 0 0 0 1  (a glider)
0 0 0 0 0 0 0 0 0 0 0 0 0 0 0 0 0 0 0 0 0 0 0 0 0 0 0 0 0 0 0 0 0 0 0
0 0 0 0 0 0 0 0 0 0 0 0 0 0 0 0 0 0 0 0 0 0 0 0 0 0 0 0 0 0 0 0 0 0 0
0 0 0 0 0 0 0 0 0 0 0 0 0 0 0 0 0 0 0 0 0 0 0 0 0 0 0 0 0 0 0 0 0 0 0
0 0 0 0 0 0 0 0 0 0 0 0 0 0 0 0 0 0 0 0 0 0 0 0 0 0 0 0 0 0 0 0 0 0 0
0 0 0 0 0 0 0 0 0 0 0 0 0 0 0 0 0 0 0 0 0 0 0 0 0 0 0 0 0 0 0 1
0 0 0 0 0 0 0 0 0 0 0 0 0 0 0 0 0 0 0 0 0 0 0 0 0 0 0 0 0 0 0 1 1
0 0 0 0 0 0 0 0 0 0 0 0 0 0 0 0 0 0 0 0 0 0 0 0 0 0 0 0 0 1 1  (a glider)
```

Figure 6.7 Gosper's glider gun

A *universal computer–constructor* is an automaton that can perform any computation (simulate any Turing machine), and which, given the description of any automaton *B*, can construct a copy of *B* in any designated empty region of the cellular space. Von Neumann was able to exhibit a universal computer–constructor, using a two-dimensional uniform cellular automaton with 29 states per cell, where any cell was directly connected to its four nearest neighbours. Self-reproduction then followed immediately since this universal computer–constructor was able to construct a copy of itself. Subsequently, many authors set out to reduce the complexity of von Neumann's machine, or even to design alternate simpler solutions. For instance, Codd [6], with remarkable ingenuity and considerable use of digital computers, showed that there exists an 8-state, 5-neighbour cellular space which is computation–construction universal.

6.3 Synchronisation problems

The most famous *synchronisation problem* studied in the theory of cellular automata is the so-called '*firing squad* synchronisation problem'. It first arose in connection with causing all parts of a self-reproducing machine to be turned on simultaneously and can be stated in the one-dimensional case as follows:

(i) Consider a finite (but arbitrarily long) one-dimensional array of synchronous finite-state machines, each of them being connected to its left and right neighbour

(ii) One of the end machines is called a 'general', and the others are called 'soldiers'.

(iii) At time $t = 0$, all the soldiers are in quiescent state and the general is in an active state labelled 'fire when ready'.

The problem is to specify the single automaton structure in such a way that after a sequence of transitions all the automata of the array enter simultaneously for the first time, the 'firing state'. The difficulty of the problem follows from the fact that the structure of the automata must be independent of the size of the array.

This problem was devised by Myhill around 1957 and first appeared in print in a paper by E. F. Moore [24]. It was first solved by J. McCarthy and M. Minsky (unpublished manuscript). Since the general must receive an answer from all the soldiers, before firing, it is easily seen that any solution needs at least $2N - 2$ steps, where N is the size of the array. A minimum-time solution was proposed by Goto [13], but the first nice solution is due to Waksman [42] who has proposed a 16-state solution which runs in time $2N - 2$ on an array of size N. Subsequently, Balzer [3] has given an alternative solution with only 8 states per automaton. The case where any cell of the array can be the general was studied by F. R. Moore and G. G.

Langdon [25]. The extension to arbitrarily connected networks has been studied by Rosenstiehl [32], Shinahr [34], Grasselli [14] and Romani [31]. The probabilistic case, that is the situation where there are random pairwise interactions between adjacent cells, has been studied by Varshavski [41].

The second synchronisation problem in the context of uniform cellular automata was introduced by Dijkstra [8] under the title 'self-stabilization in spite of distributed control'. A possible formalisation of this problem consists in defining the structure of a finite (but arbitrarily long) array of finite-state machines in such a way that, for any initial configuration:

(i) At any time t, the set of cells that are activated is represented by S_t.
(ii) For any cell i there exist infinitely many t such that i belongs to S_t.
(iii) Any machine changes its state arbitrarily often.
(iv) There is a time T such that, for any $t \geq T$, exactly one machine of the array can change its state.

Cellular structures are nice models for distributed networks, that is, structures with the following properties: there is no commonly accessible store, and therefore, the current system state must be recorded in variables distributed over the various processes; communication facilities are limited in the sense that each process can only exchange information with a small subset of processes called its neighbours.

The algorithm proposed by Dijkstra [8] may be used to guarantee for instance that at any moment only one of the machines of the array is allowed to use a shared ressource; the self-stabilisation property ensures a sort of self-repair mechanism since, in case of failure, the system automatically evolves without any external intervention towards a mutual exclusion situation.

Dijkstra [8] has exhibited a nice solution defined on a finite one-dimensional array, and which requires 4 states per cell; subsequently, Tchuente [36] has proposed a recurrent approach which yields efficient solutions for any connected network.

Example   An example from [8]:
Six cells are connected in a ring and are numbered from 0 to 5, i.e. M_i is connected to M_{i-1} and M_{i+1} where the indices are computed modulo 6. Any cell has 8 possible states: 0, 1, ..., 7. The local transition function is defined as follows:
cell M_0: if $x_6 = x_0$ then $x_0 := x_0 + 1 \pmod 6$, otherwise x_0 is unchanged
cell M_i, $i > 0$: if $x_{i-1} \neq x_i$ then $x_i := x_{i-1}$, otherwise x_i is unchanged.

An example of evolution (activated cells are italic, and those which can change state are in heavy type):

$t = 0$: 0 *1 2* 4 6 7	$t = 1$: 0 0 *2* 4 4 7	$t = 2$: 0 0 *2* 4 4 7
$t = 3$: 0 0 0 *2* 4 7	$t = 4$: 0 0 0 *2* 2 7	$t = 5$: 0 0 0 *2* 2 7
$t = 6$: 0 0 0 *2* 2 2	$t = 7$: 0 0 0 0 *2* 2	$t = 8$: 0 0 0 0 0 *2*
$t = 9$: *0 0 0 0 0 0*	$t = 10$: 1 *0 0* 0 0 0	$t = 11$: 1 1 *0 0* 0 0

6.4 Computation on uniform infinite structures

The first approach for the notion of *computation* on uniform infinite structures consists in comparing the computing capability of cellular spaces with the classical models of computation. For this approach, we shall just give a simple construction due to Smith [35], which shows how to simulate a Turing machine with a one-dimensional cellular automaton. The second approach is much more interesting from a mathematical point of view; in this model of computation, the global transition function can change during time, and the problem is to characterise the mappings Φ that can be decomposed into the form $\Phi = F_1 \circ F_2 \circ \ldots \circ F_m$, where any F_i belongs to the collection $\mathcal{F}$ of global transition functions of the network.

6.4.1 *Embedding Turing machines*

Definition A *Turing machine TM* is a quintuple (Q, X, b, q_0) where

Q is a finite set (the set of states)
X is the finite set of tape symbols
b is a particular element of X called the 'blank'
$Q \times X \mapsto X \times Q \times \{-1, 0, 1\}$ is a function
q_0 is the initial state

At any time, the configuration of a Turing machine is defined by a triple $(a_1 a_2 \ldots a_r, q, b_1 b_2 \ldots b_s)$ which specifies the current state q of the head and the symbol b_1 which is being scanned (see Fig. 6.8).
The successor of a configuration $(a_1 a_2 \ldots a_r, q, b_1 b_2 \ldots b_s)$ is

$$(a_1 a_2 \ldots a_r, q', b'_1 b_2 \ldots b_s) \qquad \text{if } d(q, b_1) = (b'_1, q', 0)$$
$$(a_1 a_2 \ldots a_r b'_1, q', b_2 \ldots b_s) \qquad \text{if } d(q, b_1) = (b'_1, q', +1)$$
$$(a_1 a_2 \ldots a_{r-1}, q', a_r b'_1 b_2 \ldots b_s) \text{ if } d(q, b_1) = (b'_1, q', -1)$$

A computation is obtained by starting *TM* in its initial state, scanning the leftmost symbol of a string $x_1 x_2 \ldots x_n$ of X^*; if the machine halts, i.e. if it reaches a configuration without any successor, then the result of the computation is the string on the tape.

Example A Turing machine that computes the sum modulo 2 of a sequence of bits can be defined as follows:

(i) $Q = \{A, B, q_0\}$, $X = \{b, 0, 1\}$,
(ii) $d(q_0, 0) = (0, A, +1)$, $d(q_0, 1) = (1, B, +1)$,
(iii) $d(A, 0) = (0, A, +1)$, $d(A, 1) = (1, B, +1)$,
(iv) $d(B, 0) = (0, B, +1)$, $d(B, 1) = (1, A, +1)$,

The machine stops when the input symbol is b; if it stops in state B then the sum $x_1 + \ldots + x_n$ (mod 2) is 1, otherwise, this sum is 0.

We are now going to show how to embed a Turing machine in a one-dimensional cellular structure [35].

In Fig. 6.8 any square of the Turing machine is associated with a cell of the cellular space. Moreover,

> if a symbol a_i is scanned by the head in state q, then the corresponding cell of the cellular automaton is in state (q, a_i)
> any other square that bears a symbol a corresponds to a cell in state $(*, a)$.

Since, during each transition, the head is stationary, or moves one square left or one square right, it is easily seen that any Turing machine can be simulated by a one-dimensional cellular automaton where any cell $i \in \mathbb{Z}$ is connected to its left and right neighbours, $i - 1$ and $i + 1$.

Comment   As noted by Smith [35], the cellular space that simulates a given Turing machine contains at most two active cells at any moment. As a consequence, the parallelism principle which led us to the study of cellular automata, is wasted when we perform computations in this way.

6.4.2   *Algebraic theory of computation on flexible structures*

In this section we are interested in the theory of computation on *flexible structures*, that is cellular automata where the transition function may change during time.

If $\aleph$ denotes the collection of permissible global transition functions of a

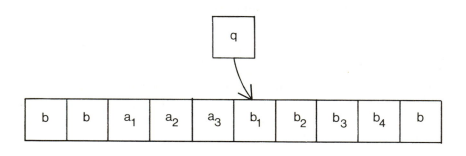

Figure 6.8

cellular structure $\aleph$, then a function Φ is said to be computable on $\aleph$ if and only if it can be decomposed into the form $\Phi = F_1 \circ F_2 \circ \ldots \circ F_m$, where any F_i belongs to $\aleph$. Given a cellular space $E = \mathbb{Z}^d$ and a finite set Q of states, let us denote by $C = \{c: E \to \mathbb{Z}\}$ the set of configurations.

The first problem is to characterise all the functions from C to C that can be realised by a cellular structure. Such functions will hereafter be called local functions. The first step towards the solution of this problem is given by the following lemma:

Lemma 1   If F and F' are local functions, then $F' \circ F$ is a local function.

Proof   Let F and F' be the global transition functions associated respectively with the cellular automata (E, Q, V', f) and (E, Q, V'', f') where $V = (v_1, \ldots, v_n)$ and $V' = (v'_1, \ldots, v'_m)$.

Let c be a configuration, and let x be a cell of E.

$$F(c)(x) = f(c(x + v_1), \ldots, c(x + v_n)) = c'(x)$$

and

$$
\begin{aligned}
F' \circ F(c)(x) &= F'(c')(x) \\
&= f'(c'(x + v'_1, \ldots, c'(x + v'_m)) \\
&= f'(f(c(x + v'_1 + v_1), \ldots, c(x + v'_1 + v_n)), \ldots, \\
&\quad f(c(x + v'_m + v_1), \ldots, c(x + v'_m + v_n)))
\end{aligned}
$$

This shows that $F' \circ F$ is the global transition function of a cellular space (E, Q, V'', f'') where any cell x is directly connected to the cells

$$x + v_i + v'_j \text{ for } 1 \leq i \leq n, \, 1 \leq j \leq m \qquad \square$$

Before coming to the statement of the theorem that characterises the local maps, let us recall some definitions.

Definition   The *shift operation* t_v associated with an element v of $E = \mathbb{Z}^d$ is defined by $t_v(x) = x + v$ for any $x \in E$. It can naturally be extended to configurations by noting $t_v(c)(x) = c(x - v)$ for any configuration c.

Hereafter, the set $C = \{c: E \to Q\}$ of configurations is identified with the product $\Pi_{x \in E} Q$. Since Q is finite, it is compact with respect to the discrete topology, and it follows that C is a compact space.

Definition   A function F from C to C is said to *commute* with the shift, if $t_v(F(c)) = F(t_v(c))$ for any configuration c.

Theorem 1   (Due to Hedlund [15]) A mapping $F: C \to C$ can be realised by a cellular automaton if and only if it is continuous and commutes with the shift.

Proof The necessary condition is obvious. For the sufficient condition, let us proceed by contradiction. So let us assume that F commutes with the shift and is continuous, but is not local. We can easily exhibit two sequences of configurations $c^{(n)}$ and $c'^{(n)}$ such that:

(i) if $B(0, n)$ denotes the cells $x = (x_1, \ldots, x_d)$ such that $|x_i| \leq n$ for any i, then the restrictions

$$c^{(n)}_{\mid B(0,n)}, \ c'^{(n)}_{\mid B(0,n)} \text{ of } c^{(n)} \text{ and } c'^{(n)} \text{ to } B(0, n) \text{ are equal.}$$

(ii) $F(c^{(n)})(0) \neq F(c'^{(n)})(0)$.

Since the set of configurations is a compact space, we can construct two subsequences $c^{(n_p)}$ and $c'^{(n_p)}$ such that

(i) $c^{(n_p)} \to c; \ c'^{(n_p)} \to c'$

(ii) $c = c'$ (since $c^{(n)}_{\mid B(0,n)} = c'^{(n)}_{\mid B(0, n)}$ for any n)

(iii) $F(c)(0) \neq F(c'(0))$

   (since $F(c^{(n)})(0) \neq F(c'^{(n)})(0)$ for any n, and Q is finite).

and this is a contradiction □

Comment This result has been extended by Richardson [29], to the nondeterministic case.

When the cellular space is fixed, an interesting question concerns the decomposition of a global map F into a finite sequence $F_p \circ \ldots \circ F_1$ of global maps associated with cellular structures with simpler neighbourhoods. This decomposition problem was first posed and partially solved by Amoroso and Epstein [1]. However, the most general result in this area is due to Nazu [27] who has established the following theorem.

Theorem 2 For any $n \geq 2$, there exist global maps of one-dimensional structures with n contiguous neighbours, that cannot be composed from a sequence of global maps associated with cellular automata with $n - 1$ contiguous neighbours.

Comment This result induces a hierarchy of cellular structures with contiguous neighbourhoods. It shows that, if the neighbourhood is enlarged, then the computation power of the cellular structure strictly increases.

More recently, Butler [5] has given for in the binary one-dimensional structures where cell i is connected to $i - 1$ and $i + 1$, an algorithm which decides whether a global map is decomposable, and if so, what decompositions are possible. A deeper analysis of decomposable maps in this particularly simple case shows that most sequences of m local maps on a two-cell neighbourhood, for large m, produce a composite map realised by a shorter sequence.

The *completeness problem* is another related question which has been introduced by Yamada and Amoroso [43], and can be stated as follows: assuming that the transition function of a cellular automaton can change over time, does there exist a finite pattern that cannot be reached from a given canonical starting pattern, by some sequence of global transformations? Usually, the starting pattern is a configuration where all cells except a single one, are in the quiescent state.

Following the partial results of Yamada and Amoroso [43] Kubo and Kimura [19] Nasu and Honda [26], Maruoka and Kimura [22] have proved the following theorem:

Theorem 3   In a one-dimensional cellular structure having a contiguous neighbourhood of size 3 or more, an arbitrary finite configuration can be reached from another arbitrary finite configuration by a finite sequence of suitably chosen global maps.

This result is very powerful, and constitutes a very good illustration of the computing power of flexible structures.

6.5   Computation on finite automata networks

In the class of infinite cellular automata, the neighbourhood relation has to be local and regular in order to permit a simple description. These conditions are no longer necessary when one deals with finite structures. So, a finite network of order n, can be defined as a triple $\mathcal{N} = (G, Q, \mathcal{F})$ where

G is a directed graph of order n representing the interconnection pattern

Q is the finite nonempty set which represents the set of states that the machines in the network can assume

$\mathcal{F}$ is a collection of functions from Q^n into itself, representing the set of possible global transition functions of the network.

Let $F = (f_1, \ldots, f_n)$ be an element of $\mathcal{F}$ where f_i is the transition function of the automaton at node i; it is easily seen that

$$\text{if } f_i(x_1, \ldots, x_n) \text{ depends on } x_j \text{ then } (j, i) \text{ is an arc of } G \qquad (6.1)$$

Any function $F = (f_1, \ldots, f_n)$ where all the components f_i verify condition (6.1) above, will be said to be *compatible* with G.

It is natural to propose the following classification of finite networks with respect to the degree of regularity of G; see Fig. 6.9.

(i) The most general class consists of arbitrary networks, that is, those where there is no constraint on the interconnection pattern.

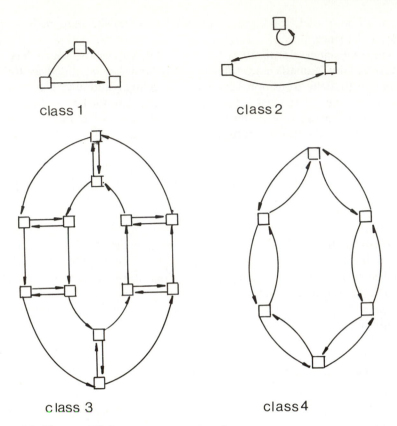

Figure 6.9 Classes of finite automaton networks

(ii) The second class consists of balanced networks defined as follows: any cell i, $1 \le i \le n$, is connected to $s_1(i)$, $\ldots$, $s_k(i)$ where s_1, $\ldots$, s_k is a fixed collection of permutations of $\{1, 2, \ldots, n\}$; the state $q^{t+1}(i)$ of cell i at time $t + 1$ is $f(q_1, \ldots, q_k)$, where q_j, $1 \le j \le k$, is the state of $s_j(i)$ (the jth neighbour of i), at time t, and f is a fixed function from Q^k to Q.

(iii) In the third class, the interconnection structure G is a Cayley graph, that is, there is a group structure on the set V of vertices, and the neighbours of any vertex v, is $\{(v, g_1 \cdot v), (v, g_2 \cdot v), \ldots, (v, g_m \cdot v)\}$, where $g_1, \ldots, g_m$ is a fixed collection of elements of V. The transition functions are defined as in the second class. Such networks are said to be uniform.

(iv) The fourth class is a subset of the third class, and is defined by considering interconnection structures that are Cayley graphs associated with communtative groups. Such networks are also called arrays. In this context one can study the possibility of one network simulating another network, in order to detect network similarities. They have shown that any balanced network can be realised by a uniform network, and that there are balanced networks that cannot be realised by arrays. As a consequence, the class of arrays is less powerful than the class of balanced networks and the class of uniform networks.

Many problems that were intractable for infinite cellular structures can be solved now, and they often require the introduction of nice mathematical tools.

6.5.1 Computation on flexible networks

In this section, $A(Q^n)$, $\text{Sym}(Q^n)$ and $\text{Alt}(Q^n)$ denote respectively the set of mappings from Q^n into itself, the symmetric group over Q^n and the alternate group over Q^n. In networks of the form $\mathcal{N} = (G, Q, A(Q^n))$, that is, where the constraints on the local transition functions are only due to the interconnection pattern, the computable functions are characterised by the following theorem [40].

Theorem 4 For any finite set Q of cardinality greater than one, $A(Q^n)$ is computable on a network $\mathcal{N} = (G, Q, A(Q^n))$ of order n if and only if G is strongly connected and contains a vertex v_r such that, for any vertex $v_i \neq v_r$, (v_i, v_r) is an arc of G.

Proof Let us identify Q with the ring of integers modulo q, where $q = |Q|$.

Sufficient condition: From the hypothesis, any function of the form

$$E_{a,r}(x) = \begin{cases} a + e_r & \text{if } x = a \\ a & \text{if } x = a + e_r \\ x & \text{otherwise} \end{cases}$$

or

$$E'_{a,r}(x) = \begin{cases} a + e_r & \text{if } x = a \\ x & \text{otherwise} \end{cases}$$

is compatible with G, and is thus computable on $\mathcal{N}$.

On the other hand, since G is strongly connected, it follows from Gastinel [11], Tchuente [37], Kim and Roush [17], that any linear mapping is computable on $\mathcal{N}$. It is then an easy matter to show that any mapping of the form

$$E_{a,j}(x) = \begin{cases} a + e_j & \text{if } x = a \\ a & \text{if } x = a + e_j \\ x & \text{otherwise} \end{cases}$$

or

$$E'_{a,j}(x) = \begin{cases} a + e_j & \text{if } x = a \\ x & \text{otherwise,} \end{cases}$$

where e_j denotes the jth element of the canonical basis of Q^n is computable on $\mathcal{N}$. Since the collection of functions $\{E_{a,j}, E'_{a,j}; a \in Q^n, 1 \leq j \leq n\}$ generates $A(Q^n)$, it follows that $A(Q^n)$ is computable on $\mathcal{N}$.

Necessary condition: It is obvious that G must be strongly connected. Now, let $F \in A(Q^n)$ be such that

$$|\mathrm{Im}(F)| = q^n - 1, \text{ where } \mathrm{Im}(F) = \{F(x): x \in Q^n\}$$

It is easily seen that, if $F = F_p \circ \ldots \circ F_1$, then there is an index i such that $|\mathrm{Im}(F_i)| = q^n - 1$. It is therefore sufficient to show that such a function cannot be a transition function of a network whose graph G does not satisfy the conditions of the theorem.

Any function F such that $|\mathrm{Im}(F)| = q^n - 1$ verifies

$$\Sigma_{x \in Q^n} F(x) = b - a \text{ where } |F^{-1}(a)| = \emptyset \text{ and } |F^{-1}(b)| = 2 \qquad (6.2)$$

Let $F = (f_1, \ldots, f_n)$ be the transition function of a network where no vertex satisfies the condition of the theorem. If f_i depends for instance on $x_1, \ldots, x_p$ where $p < n$, then, by denoting

$$J = \{x \in Q^n: x_k = 0 \text{ for } k = p + 1, \ldots, n\},$$

it is easily seen that, for any $y \in Q$

$$|f_i^{-1}(y)| = q^{n-p} \cdot |f_i^{-1}(y) \cap J| = 0 \pmod{q}$$

As a consequence,

$$\Sigma_{x \in Q^n} F(x) = 0 \pmod{q} \qquad (6.3)$$

The theorem follows from the contradiction between (6.2) and (6.3)   □

Examples   are shown in Fig. 6.10.

A graph is called a *tree*, if

It is symmetric: if (i, j) is an arc of G, then (j, i) is also an arc of G
It does not contain a circuit of length greater than 2

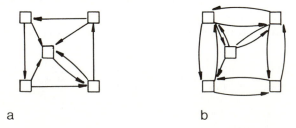

a   b

Figure 6.10 (a) verifies the conditions of the theorem; (b) does not verify the conditions of the theorem

In the case where the interconnection graph G of the network is a tree, we say that $\mathcal{N} = (G, Q, A(Q^n))$ is a tree-network. See Fig. 6.11.

The following result concerns the computation of $\mathrm{Sym}(Q^n)$ and $\mathrm{Alt}(Q^n)$ on tree-networks.

Proposition 1   Let $\mathcal{N} = (G, Q, A(Q^n))$ be a tree-network.

(i) If $Q = \{0, 1\}$, then $\mathrm{Alt}(Q^n)$ is computable on $\mathcal{N}$ and $\mathrm{Sym}(Q^n)$ is computable on $\mathcal{N}$ if and only if G is a star
(ii) If $|Q| > 2$, $\mathrm{Alt}(Q^n)$ is computable on $\mathcal{N}$.
(iii) If $|Q| > 2$ is odd, then $\mathrm{Sym}(Q^n)$ is computable on $\mathcal{N}$.

   In order to characterise the set of all mappings computable on a given binary tree-network, we need some definitions.

Definition   In a tree, a vertex of degree one is said to be *pending*. A *caterpillar* is a tree obtained from a chain by adding pending vertices (see Fig. 6.12).

We are now ready to state the main theorem:

Theorem 5   A function F from $\{0, 1\}^n$ to itself is computable on a binary tree-network associated with a tree H if and only if

$$\Sigma_{x \in Q^n}[|F^{-1}(x)|/2] \geq 2^{n-p-1}$$

where p is the maximum number of pending vertices of a caterpillar of H.

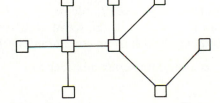

Figure 6.11 A tree network

a chain   a caterpillar

Figure 6.12 Trees, showing pending vertices darkened

Proof The proof of this result is quite lengthy, and may be found in [40].

Comment This result shows that, as far as the computation of boolean functions is concerned, there is a total preorder between trees. More precisely, the trees of order n can be partitioned into $n - 2$ classes $C_2, C_3, \ldots$ C_{n-1}, where C_i consists of the trees whose caterpillars contain at most i pending vertices. The theorem says that if $\mathcal{N} = (H, Q, A(Q^n))$ and $\mathcal{N}' = (H', Q, A(Q^n))$ where $Q = \{0, 1\}$, $H \in C_i$ and $H' \in C_{i'}$ with $i < i'$, then any function computable on $\mathcal{N}$ is also computable on $\mathcal{N}'$.

It is important to note that this hierarchy does not depend on the maximum degree of a vertex in G. For instance, in the example of Fig. 6.13, the binary tree-network $\mathcal{N}' = (H', Q, A(Q^{12}))$, $Q = \{0, 1\}$ computes more functions than $\mathcal{N}$.

The computation of *monotone boolean functions* is quite difficult. So let us now consider rather the computation of *boolean matrices*. Clearly, a square boolean matrix A of order n is compatible with a directed graph G if $a_{ij} = 0$ for all couples (i, j) such that (j, i) is not an arc of G.

If A is a square boolean matrix of order n, we denote $A(i) = \{j: a_{ij} = 1\}$, $1 \le i \le n$. A square boolean matrix A of order n is said to be *prime* if, for any factorisation $A = B \cdot C$ where B, C are square boolean matrices of order n, B or C is a boolean permutation matrix.

It is easily seen that, if A is not a boolean permutation matrix, then, in any factorisation $A = A_1 A_2 \ldots A_p$, there exists an index i such that A_i is not a boolean permutation matrix. This shows that prime boolean matrices play a central role in the study of the computation of boolean matrices.

For any tree H of order n, there exist prime boolean matrices compatible with H and characterised by the following theorem [37].

Theorem 6 Let $H = (V, E)$ be a tree of order n, where $V = \{1, 2, \ldots, n\}$ is the set of vertices and $E \subseteq V \times V$ is the set of arcs. Any square boolean matrix of order n verifying the following conditions is prime.

(i) $A(i) = \{i, j\}$ if i is a pending vertex adjacent to j

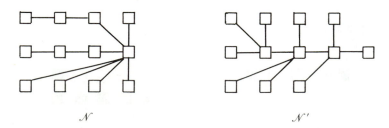

Figure 6.13 Comparison of tree-networks

(ii) $A(i) = \{j: j \neq i \text{ and } [j, i] \in E\}$ if i is adjacent to a pending vertex
(iii) in all other cases

$$A(i) = \{j: j \neq i, [j, i] \in E\} \text{ or } A(i) = \{j: j = i \text{ or } [j, i] \in E\}$$

Let $\mathcal{M}_n$ denote the set of square boolean matrices of order n. This theorem implies that, for any network $\mathcal{N} = (H, \{0, 1\}, \mathcal{M}_n)$ and where H is a tree, there exist boolean matrices of order n which are computable on $\mathcal{N}$, but which cannot be computed on any network $\mathcal{N}' = (H', \{0, 1\}, \mathcal{M}_n)$ where H' is a tree of order n different from H.

As a consequence, binary tree-networks are mutually uncomparable with respect to the computation of boolean matrices. This situation is to be compared with the hierarchy established previously among trees, with respect to the computation of general boolean functions.

Let us now turn to the study of the *factorisation lengths*. When $F = F_p \circ \ldots \circ F_1$, where the F_i are global transition functions of a network $\mathcal{N}$, the length p of the *factorisation* is the time necessary to compute F on $\mathcal{N}$. An interesting problem is thus to minimise p. More precisely we consider the factorisation of F with minimum length,

$$F = F_{p*} \circ \ldots \circ F_1$$

and we denote $p* = l_{\mathcal{N}}(F)$. Clearly, if F is computable on $\mathcal{N} = (G, Q, \mathcal{F})$, and $\mathcal{N}' = (G', Q, \mathcal{F})$, where G' contains G, then $l_{\mathcal{N}}(F) \leq l_{\mathcal{N}'}(F)$

An interesting question is the determination of an optimal structure for the computation of a given family of functions. Because of the preceding remark, the problem must be posed only for a class of networks where all the graphs have the same number of arcs. A typical example of such graphs is the class of trees.

On the other hand, the computation of a function

$$F_\sigma(x_1, \ldots, x_n) = (x_{\sigma(1)}, \ldots, x_{\sigma(n)}),$$

where σ is a permutation, can model a situation where, for $i = 1, \ldots, n$, the automaton at node i wants to send a message to the automaton at node $\sigma(i)$. The minimum factorisation length $l_{\mathcal{N}}(F)$ can therefore be considered as the delay necessary to transmit in a network $\mathcal{N}$, the collection of messages $(i, \sigma(i))$, $i = 1, \ldots, n$. Clearly, if the automata of the network are only allowed to receive and transmit one message at a time, and without any modification, then any function F_σ must be factorised into a product of boolean permutation matrices.

We are now going to compare trees with respect to this communication criteria [37].

Proposition 2 If H is the chain of order $n \geq 3$, then $l_H = n$.

Proof Let us first note that an algorithm which sorts a sequence corresponding to a permutation σ consists in applying the permutation σ^{-1} to the given sequence; as a consequence, sorting a permutation σ is equivalent to realising the permutation σ^{-1}.

An inversion of a permutation σ, is a couple $(\sigma(i), \sigma(j))$ such that $i < j$ and $\sigma(i) > \sigma(j)$. It is easily verified that:

(i) The identity permutation contains no inversion
(ii) An exchange between two adjacent vertices of a chain increases or decreases the number of inversions by one.

Let us now show that $l_H \geq n$. The permutation defined by

$$\sigma(i) = n - i + 1 \text{ for } i = 1, \ldots, n$$

contains $n(n - 1)/2$ inversions. Thus its realisation on the chain needs exactly $n(n - 1)/2$ exchanges.
If n is odd, then one can perform at most $(n - 1)/2$ simultaneous exchanges on the chain of order n, thus $l_H \geq n$.
If n is even and if at some step, the number of inversions is decreased by $n/2$, then the resulting permutation is of the form

$$a(1) < a(2), a(3) < a(4), \ldots, a(n - 1) < a(n)$$

Therefore, at the next step, the number of inversions can be decreased by at most $n/2 - 1$; this shows that $l_H \geq n$.

In order to show that $l_H \leq n$, let us consider the algorithm that alternately performs the following steps:

(i) Perform simultaneously the exchanges between adjacent vertices $(2i - 1, 2i)$, in order to obtain a permutation a such that

$$a(1) < a(2), a(3) < a(4), \ldots, a(2i - 1) < a(2i), \ldots$$

(ii) Perform simultaneously the exchanges between adjacent vertices $(2i, 2i + 1)$, in order to obtain a permutation a such that

$$a(2) < a(3), a(4) < a(5), \ldots, a(2i), < a(2i + 1), \ldots$$

This algorithm sorts any permutation of order n, in n steps or less (see Knuth [18]).

Example

```
5 = = = = = 3 − − − − − 4 = = = = = 1 − − − − − 2
3 − − − − − 5 = = = = = 1 − − − − − 4 = = = = = 2
3 = = = = = 1 − − − − − 5 = = = = = 2 − − − − − 4
1 − − − − − 3 = = = = = 2 − − − − − 5 = = = = = 4
```

$$1 = = = = = 2 - - - - - 3 = = = = = 4 - - - - - 5$$
$$1 - - - - - 2 - - - - - 3 - - - - - 4 - - - - - 5$$

Before coming to the analysis of arbitrary trees, let us recall some notations and definitions. The connected components of the forest obtained from a tree H by removing a vertex i and its incident edges, are called the *branches* at i. The size of a branch is the number of its vertices.

In [16], Jordan has shown that any tree H of order n admits a centre of the first kind, that is, a vertex c whose branches are of maximum size $q \leq n/2$; moreover, if $q < n/2$, then c is unique, otherwise H admits exactly one other centre c' which is adjacent to c. The following lemma is from [39]:

Lemma 2   If a tree $H = (V, E)$ of order n admits a unique centre c, then there exists a permutation σ over V such that $\sigma(c) = c$ and, for any $v \neq c$, v and $\sigma(v)$ do not belong to the same branch at c.

Proof   If H is a tree of order $n \leq 3$, then the result is trivial, otherwise let $H' = (V', E')$ be the graph defined by

(i) $V' = V - \{c\}$,
(ii) $[v, v'] \in E'$ if and only if v and v' do not belong to the same branch at c.

Since the branches at c are of maximum size $q < n/2$, it is easily seen that

$$d_{H'}(v) \geq n - 1 - q \geq |V'|/2 \text{ for any } v \in V' \text{ (where } |V'| \geq 3)$$

Hence H' is Hamiltonian [9], and the result holds since we just have to consider a permutation associated with any Hamiltonian cycle of H'.   □

We are now ready to state the general theorem on the comparison of trees with respect to the factorisation of boolean permutation matrices [37].

Theorem 7   If $\mathcal{N} = (G, Q, \mathcal{F})$ where G is a tree of order n, and $\mathcal{F}$ is the class of boolean permutation matrices, then

$$\max_{\sigma \in S_n} l_{\mathcal{N}}(F_\sigma) \geq n,$$

and this bound is tight if G is the chain.

Proof   Every exchange involving a given vertex moves exactly two objects, and two of them cannot be performed simultaneously. Thus, if an algorithm that realises a permutation σ over V in a minimum number p^* of steps, performs $n(i, v)$ exchanges involving element i and vertex v, then

$$p^* \geq \max_{v \in V}(\Sigma_i n(i, v))/2$$

Case 1: G admits a unique centre c.
Let σ be one of the permutations exhibited in the preceding lemma. It is easily seen that $n(i, c) \geq 2$ for any element i. Thus

$$p^* \geq (\Sigma_i n(i, c)/2 \geq n$$

Case 2: G admits two centres c, c'; see Fig. 6.14.

If B (resp. B') denotes the branch at c' (resp. c) containing c (resp. c'), then $|B| = |B'| = n/2$. Hence there exists à permutation σ over V such that

(i) $\sigma(c) = c'$, $\sigma(c') = c$,

(ii) if i is a vertex of B, then $\sigma(i)$ is a vertex of B' (and conversely).

If the first exchange involving vertex c is performed with a vertex of B then

$$n(c', c) \geq 3, n(c, c) \geq 1, \text{ and } n(i, c) \geq 2 \text{ if } (i \neq c \text{ and } i \neq c')$$

Hence

$$p^* \geq \Sigma_i n(i, c))/2 \geq n$$

If the first exchange involving vertex c is performed with vertex c' of B' then

$$n(c', c') \geq 3, n(c, c') \geq 1, \text{ and } n(i, c') \geq 2 \text{ if } (i \neq c \text{ and } i \neq c')$$

hence

$$p^* \geq \Sigma_i n(i, c'))/2 \geq n$$

This ends the proof of the theorem □

Let us now turn to the study of minimum length factorisations, when the set of global transition functions is enlarged. Clearly, if F is computable on $\mathcal{N} = (G, Q, \mathscr{F})$, and $\mathcal{N}' = (G, Q, \mathscr{F}')$, where $\mathscr{F} \subseteq \mathscr{F}'$, then $l_{\mathcal{N}'}(F) \leq l_{\mathcal{N}}(F)$

Example

The graph G is the chain of order 3, and $F(x_1, x_2, x_3) = (x_3, x_2, x_1)$.

(i) If $\mathscr{F}$ is the set of boolean permutation matrices of order 3 then $l_{\mathcal{N}}(\mathscr{F}) = 3$, and the corresponding factorisation is $F = F_3 \circ F_2 \circ F_1$, where

$$F_1(x) = (x_2, x_1, x_3), F_2(x) = (x_1, x_3, x_2), F_3 = F_1$$

(ii) If $\mathscr{F}'$ is the set of linear transformations, then $l_{\mathcal{N}'}(F) = 2$, and the corresponding factorisation is $F = F'_2 \circ F'_1$ where

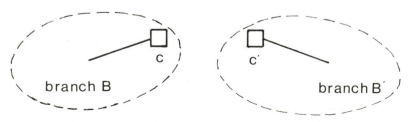

Figure 6.14 Case with two centres

$$F'_1(x) = (x_1 + x_2, x_1 + x_2 + x_3, x_2 + x_3),$$
$$F'_2(x) = (x_2 - x_1, x_1 - x_2 + x_3, x_2 - x_3)$$

We are now going to establish a universal bound on the number of steps necessary for the transmission of a collection of n binary messages within a star-connected network of n binary automata. In order to prove this result, we need some preliminary lemmas.

Lemma 3 If σ is a cyclic permutation of the form $\sigma = (1, a_2, \ldots, a_k)$ or $\sigma = (a_2, a_3, \ldots, a_k)$ with $\sigma(1) = 1$, then $l_{\mathcal{N}}(F_\sigma) \leq k - 1$.

Proof

Case 1: $\sigma = (1, a_2, \ldots, a_k)$. Clearly,

$$F_\sigma = F_{s^{(k)}} \circ \ldots \circ F_{s^{(2)}}$$

where any $s^{(i)} = (1, a_i)$, $i = 2, \ldots, k$, is compatible with H. Hence $l_{\mathcal{N}}(F_\sigma) \leq k - 1$

Case 2: $\sigma = (a_2, a_3, \ldots, a_k)$ with $\sigma(1) = 1$. Let

$$F_1(x_1, \ldots, x_n)$$
$$= (x_1 + x_2 + x_k, x_1 + x_2, x_3, \ldots, x_{k-1}, x_1 + x_k, x_{k+1}, \ldots, x_n)$$

$$F_2(x_1, \ldots, x_n) = (x_3 + x_k, x_1 - x_2, x_1 + x_3, x_4, \ldots, x_n)$$

$$F_i(x_1, \ldots, x_n) = (x_{i+1} + x_k, x_2, \ldots, x_{i-1}, x_i - x_1, x_1 + x_{i+1}, x_{i+2}, \ldots, x_n),$$
$$i \geq 3$$

It is easily verified that for $2 \leq i \leq k - 2$,

$$F_i \circ F_{i-1} \circ \ldots \circ F_1(x)$$
$$= (x_1 + x_{i+1} + x_k, x_k, x_k, x_2, \ldots, x_{i-1}, x_1 + x_i + x_{i+1}$$
$$+ x_k, x_{i+2}, \ldots, x_{k-1}, x_1 + x_k, x_{k+1}, \ldots, x_n)$$

Hence

$$F_{k-2} \circ F_{k-3} \circ \ldots \circ F_1(x)$$
$$= (x_1 + x_{k-1} + x_k, x_k, x_2, \ldots, x_{k-3}, x_1 + x_{k-2} + x_{k-1} + x_k,$$
$$x_1 + x_k, x_{k+1}, \ldots, x_n)$$

Clearly, if $F_{k-1}(x) = (x_k - x_2, x_2, \ldots, x_{k-2}, x_{k-1} - x_1, x_1 - x_k, x_{k+1},$ $\ldots, x_n)$, then $F_{k-1} \circ F_{k-2} \circ \ldots \circ F_1(x) = F_\sigma$, hence $l_{\mathcal{N}}(F_\sigma) \leq k - 1$. $\square$

Lemma 4 For any permutation σ of order n, $l_{\mathcal{N}}(F_\sigma) \leq n - 1$

Proof (follows directly from the preceding lemma).

(i) If $\sigma = (1, a_2, \ldots, a_r)(b_1, b_2, \ldots, b_s) \ldots (z_1, z_2, \ldots, z_t)$, then

$$l_{\mathcal{N}}(F_\sigma) \leq l_{\mathcal{N}}(F_{(1,a_2,\ldots,a_r)}) + l_{\mathcal{N}}(F_{(b_1,\ldots,b_s)}) + \ldots + l_{\mathcal{N}}(F_{(z_1,\ldots,z_t)})$$
$$\leq (r - 1) + s + \ldots + t \leq n - 1$$

(ii) If $\sigma = (a_1, \ldots, a_r)(b_1, b_2, \ldots, b_s) \ldots (z_1, z_2, \ldots, z_t)$ with $\sigma(1) = 1$, then

$$l_{\mathcal{N}}(F_\sigma) \leq l_{\mathcal{N}}(F_{(a_1,\ldots,a_r)}) + l_{\mathcal{N}}(F_{(b_1,\ldots,b_s)}) + \ldots + l_{\mathcal{N}}(F_{(z_1,\ldots,z_t)})$$
$$\leq r + s + \ldots + t \leq n - 1$$

This ends the proof of the lemma.   □

Let σ be a permutation of order n. Since F_σ is bijective, it follows that any factorisation of F_σ is composed of bijective mappings. Therefore, in order to obtain a lower bound for $l_{\mathcal{N}}(F_\sigma)$, we are going to study the structure of mappings F of $A(\{0, 1\}^n)$ which are bijective and compatible with the star. Any mapping f from $\{0, 1\}^n$ to $\{0, 1\}$ is identified with its unique reduced representative polynomial over the ring of integers modulo 2.

Lemma 5   See [38]. If $F = (f_1, \ldots, f_n)$ is bijective and compatible with the star, then any f_j, $i \geq 2$ is affine (we assume that $i = 1$ for the centre cell).

Proof   For any $i \geq 2$, f_i can depend only on x_i and x_1, and its reduced representative polynomial is of the form

$$P_i(x_1, x_i) = a + bx_1 + cx_i + dx_1x_i \text{ where } a, b, c, d \in \{0, 1\}$$

If $d = 1$ then $|f_i^{-1}(0)| \neq 2^{n-1}$ and this contradicts the fact that F is bijective. As a consequence $d = 0$ and f_i is affine.   □

Lemma 6   See [39]. If $\sigma = (a_2, a_3, \ldots, a_n)$, $\sigma(1) = 1$, then $l_{\mathcal{N}}(F_\sigma) \geq n - 1$.

Proof   Let $F_\sigma = F_q \circ \ldots \circ F_1$, where $F_i = (f_{i,1}, \ldots, f_{i,n})$ for $i = 1, \ldots, q$, be a factorisation of F_σ with minimum length, and let us denote

$$F_j \circ \ldots \circ F_1 = (g_{j,1}, \ldots, g_{j,n}), \text{ for } j = 1, \ldots, q$$

Since from the preceding lemma any $f_{i,k}$ is affine, it is easily verified that any $g_{j,k}$ with $1 \leq j \leq q$, $2 \leq k \leq n$ is of the form

$$g_{j,k} = c_{j,k}x_1 + d_{j,k}x_k + \Sigma_{1 \leq i < q} a_{j,k}^{(i)} g_{i,1}.$$

Since $F_q \circ \ldots \circ F_1 = (g_{q,1}, \ldots, g_{q,n}) = F_\sigma$, we can write

$$u_2 = x_n - c_{q,2}x_1 - d_{q,2}x_2 = \Sigma_{1 \leq i < q} a_{q,2}^{(i)} g_{i,1}$$
$$u_3 = x_n - c_{q,3}x_1 - d_{q,3}x_3 = \Sigma_{1 \leq i < q} a_{q,3}^{(i)} g_{i,1}$$
$$\vdots$$
$$u_n = x_n - c_{q,n}x_1 - d_{q,n}x_n = \Sigma_{1 \leq i < q} a_{q,n}^{(i)} g_{i,1}$$

In the vector space of polynomials over $\{0, 1\}$, the elements $\{u_2, \ldots, u_n\}$, when decomposed with respect to the independent system $\{x_1, x_2, \ldots, x_n\}$, yield the matrix

$$M = \begin{bmatrix} -c_{q,2} & -c_{q,3} & & & -c_{q,n} \\ -d_{q,2} & 1 & & & \\ & -d_{q,3} & 1 & & \\ & & & 1 & \\ 1 & & & & -d_{q,n} \end{bmatrix}$$

and it is easily verified that $\text{rank}(M) \geq n - 2$. Since any u_i is generated by $\{g_{i,1}, 1 \leq i \leq q\}$, it follows that $q - 1 \geq n - 2$; hence $l_{\mathcal{N}}(F_\sigma) = q \geq n - 1$.

$\square$

All these results can be summarised in the following theorem [39]:

Theorem 8 If $\mathcal{N} = (G, Q, \mathcal{F})$, where G is a star of order n, and $\mathcal{F}$ is the set of functions from $\{0, 1\}^n$ into itself, then

$$\max_{\sigma \in S_n} l_{\mathcal{N}}(F_\sigma) = n - 1.$$

Comment It is important to note that the diameter of a star of order n is 2. As a consequence, the bound $n - 1$ is due to the bottleneck that occurs at the central node of a star, when several messages must travel through a star-connected network. In the theorem, any boolean function can be applied to a collection of boolean messages that have to be transmitted, but the proof shows that optimal algorithms are obtained with linear transformations.

6.5.2 *Other models of computation*

The *iterative model of computation* is a model where a property (P) is said to be computable on $\mathcal{N} = (G, Q, \mathcal{F})$ if there exists a function F of $\mathcal{F}$ such that, for any initial state $x = (x_1, \ldots, x_n)$, the network performs a stationary iteration

$$x, F(x), \ldots, F^m(x) = F^{m+1}(x)$$

and $F^m(x)$ displays property (P).

This model is well suited to the solution of graph problems. For instance, if (P) is k-colourability, then $F^m(x)$ must exhibit a k-colouration of the interconnection graph G associated with the network.

The main results obtained in this area can be summarised as follows [4]:

Theorem 9 There exist finite automata networks for the following graph problems:

connectivity
is G a maximal bichromatic graph?
computation of a rooted tree

computation of an Eulerian cycle
computation of a Hamiltonian cycle

The drawback of this approach is that, when a property (P) is investigated for a graph H of order n, the interconnection structure G of the network that solves the problem must contain H.

From a practical point of view, the most powerful model of computation for finite automata networks, is the model of systolic arrays. Systolic arrays, as defined by Kung and Leiserson [21], are a useful tool for special-purpose VLSI system design. Typically a systolic network of processors is implemented with only a few types of elementary cells interconnected in a simple, local and regular way. This makes them economical to manufacture and repair. Moreover, in a context where the network consists of single-chip microprocessors used in groups of tens or hundreds, the system can be enlarged to accommodate more variables by simply adding more chips. The high performance achieved by systolic arrays is obtained by extensive use of multiprocessing and pipelining.

It is important to note that, contrary to the classical pipeline scheme where pipelining is applied only to result variables (see the variables c_{ij} in Fig. 6.15), the two-dimensional systolic algorithms also pipeline data-variables (see the variables a_{ik}, b_{kj} in Fig. 6.16 which is based on [23]). In the figures, we illustrate the computation of the product $C = A \cdot B$ of two $n \times n$ matrices A and B on a rectangular $n \times m$ array. The elementary cells of the array operate as follows:

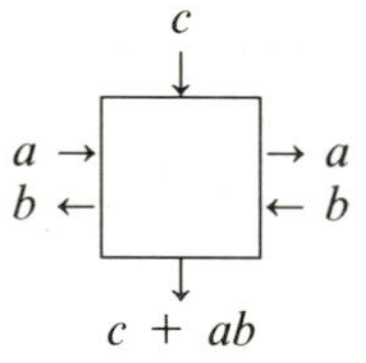

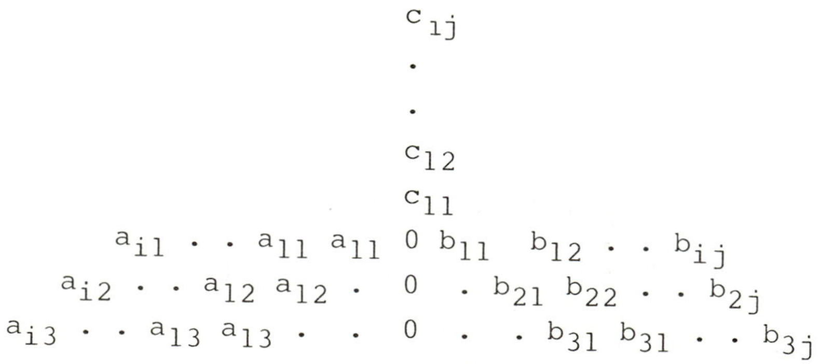

Figure 6.15 Classical pipeline scheme – a partial view of the algorithm; with $n = 3$, $m = 1$

$$c_{12} \quad c_{13} \quad c_{23}$$
$$c_{11} \quad c_{22} \quad c_{33}$$
$$c_{21} \quad c_{31} \quad c_{32}$$
$$\cdot \qquad \cdot \qquad \cdot$$
$$\cdot \qquad \cdot \qquad \cdot$$

$$a_{11} \ a_{11} \ a_{21} \ a_{31} \ a_{31} \ \ 0 \ \ \ 0 \ \ \ 0 \ b_{11} \ b_{11} \ b_{12} \ b_{13} \ b_{13}$$
$$a_{12} \ a_{12} \ a_{22} \ a_{32} \ a_{32} \ \cdot \ \ 0 \ \ \ 0 \ \ \ 0 \ \cdot \ b_{21} \ b_{21} \ b_{22} \ b_{23} \ b_{23}$$
$$3 \ a_{13} \ a_{23} \ a_{33} \ a_{33} \ \cdot \ \cdot \ \ 0 \ \ \ 0 \ \ \ 0 \ \cdot \ \cdot \ b_{31} \ b_{31} \ b_{32} \ b_{33} \ b_{33}$$

Figure 6.16 Two-dimensional systolic algorithm; with $n = m = p = 3$

This pipelining of data variables enables multiple computations to be performed for every I/O access. As a consequence, as explained in [20], systolic designs lead to structures that can solve compute-bound problems without unreasonably increasing I/O requirements.

A detailed presentation of systolic arrays is given in Chapter 8.

6.6 References

[1] S. Amoroso and I. J. Epstein, Indecomposable parallel maps of tessellation structures, *J. Comput. Syst. Sci.*, 6 (1976), 136–42
[2] M. A. Arbib, *Theories of Abstract Automata*, Prentice-Hall, Englewood Cliffs, NJ (1969)
[3] R. Balzer, An 8-state minimal time solution to the Firing Squad Synchronization Problem, *Inf. & Control*, 10 (1967), 22–42
[4] J. Berstel, Quelques applications des réseaux d'automates à des problèmes de la théorie des graphes, Thesis, Paris (1967)
[5] J. T. Butler, Synthesis of one-dimensional binary cellular structures from composite local maps, *Inf. & Control*, 43, 3 (1979), 304–24
[6] E. F. Codd, *Cellular Automata*, Academic Press, New York (1968)
[7] J. Conway, Mathematical games, *Scientific American*, (1970), 120–3 (paper written by M. Gardner)
[8] E. W. Dijkstra, Self-stabilizing systems in spite of distributed control, *Comm. ACM*, 17, 11 (1974), 643–4
[9] P. Erdös and T. Gallai, On maximal paths and circuits of graphs, *Acta Math. Acad. Sci. Hung.*, 10 (1959), 337–56
[10] F. Fogelman Soulié, Contribution à une théorie du calcul sur réseaux, Thesis, Grenoble (1985)
[11] N. Gastinel, Réalisation du calcul d'une transformation linéaire aux noeuds d'un graphe, *Colloque sur les méthodes de calcul pour des systèmes de type coopératif*, Giens, France (1978)
[12] E. Goles Chacc, Comportement dynamique de réseaux d'automates, Thesis, Grenoble (1985)
[13] E. Goto, A minimal-time solution to the firing squad problem, Course notes for applied mathematics, Harvard University Cambridge, Mass. (1962) 52–9

[14] A. Grasselli, Synchronization of cellular arrays: the firing squad synchronization problem in two dimensions, *Inf. & Control*, 28 (1975), 113–24

[15] G. A. Hedlund, Endomorphisms and automorphisms of the shift dynamical system, *Math. System Theory*, 3 (1969), 320–75

[16] C. Jordan, Sur les assemblages de lignes, *J. Reine Angew. Math. Crelle*, 70 (1869), 185–90

[17] K. H. Kim and F. W. Roush, Realizing all linear transformations, *Linear Algebra and its Applications*, 37 (1981), 97–101

[18] D. E. Knuth, *The Art of Computer Programming* Vol. 3, Addison-Wesley Reading, Mass. (1973)

[19] Kubo and Kimura, On completeness problems of tessellation automata, *IECE*, October (1972), 72–9

[20] H. T. Kung, Why Systolic architectures, *Computer Magazine*, 15, 1 (1982), 37–46

[21] H. T. Kung and C. E. Leiserson, Systolic arrays for VLSI, in *Introduction to VLSI Systems*, C. A. Mead and L. A. Conway (eds) Addison-Wesley, Reading, Mass. (1980), Sect. 8.3, 37–46

[22] A. Maruoka and M. Kimura, Decomposition phenomenon in one-dimensional scope-three tessellation automata with arbitrary number of states, *Inf. & Control*, 34 (1977), 296–313

[23] L. Melkemi and M. Tchuente, Programmation du produit matriciel sur un réseau systolique rectangulaire, *TSI* 4, 5 (1985), 459–69

[24] E. F. Moore, The firing squad synchronization problem, in *Sequential Machines*, E. F. Moore (ed.) Addison-Wesley, Reading, Mass. (1964), 213–14

[25] F. R. Moore and G. G. Langdon, A generalized firing squad problem, *Inf. & Control*, 12 (1968), 212–20

[26] Nasu and Honda, A completeness property of one-dimensional tessellation automata, *J. Comput. & Syst. Sci.*, 12 (1976), 36–48

[27] M. Nazu, Indecomposable local maps of tessellation automata, *Math. Syst. Theory*, 13 (1979), 81–93

[28] J. von Neumann, *Theory of Self-Reproducing Automata*, A. W. Burks (ed.), University of Illinois Press, Urbana (1966)

[29] D. Richardson, Tessellation with local transformation, *J. Comput. & Syst. Sci.*, 6 (1972), 373–88

[30] F. Robert, *Discrete Iterations, a metric study*, Springer Series in Computational Mathematics, Springer Verlag, Berlin, Heidelberg, New York (1986)

[31] F. Romani, Cellular automata synchronization, *Information Sciences*, 10 (1976), 299–318

[32] P. Rosenstiehl, Existence d'automates d'états finis capables de s'accorder bien qu'arbitrairement connectés et nombreux, *International Computation Centre Bulletin*, 5 (1966), 215–44

[33] P. Rosenstiehl, R. Fiskel and A. Hollinger, Intelligent graphs: networks of automata capable of solving graph problems, in *Graph Theory and Computing*, R. C. Read (ed.) Academic Press, New York (1973), 219–65

[34] I. Shinahr, Two and three-dimensional firing squad synchronization problems, *Inf. & Control*, 24 (1974), 163–80

[35] A. R. Smith III, Cellular automata theory, Tech. Rep. 2, Stanford University, Calif. (1969)

[36] M. Tchuente, Sur l'auto-stabilisation dans un réseau d'ordinateurs, *RAIRO Theoretical Computer Science*, 15, 1 (1981), 47–66

[37] M. Tchuente, Contribution à l'étude des méthodes de calcul pour des systèmes de type coopératif, Thesis, Grenoble (1982)

[38] M. Tchuente, Computation of boolean functions on networks of binary

automata, *J. Comput. & Syst. Sci.*, 26, 2 (1983), 269–77
[39] M. Tchuente, Permutation factorization on star-connected networks of binary automata, *Siam J. Alg. Disc. Meth.* 6, 3 (1985), 537–40
[40] M. Tchuente, Computation on binary tree-networks, *Discrete Applied Math.*, 14 (1986), 295–310
[41] V. I. Varshavsky, Synchronization of a collection of automata with random pairwise interaction, *Automata Remote Contr.*, 29 (1969), 224–8
[42] A. Waksman, An optimum solution to the firing squad synchronization problem, *Inf. & Control*, 9 (1966), 66–78
[43] H. Yamada and S. Amoroso, A completeness problem for pattern generation in tessellation automata, *J. Comput. & Syst. Sci.*, 4 (1970), 137–76

Part 2
Applications

7 *Françoise Fogelman Soulié, Patrick Gallinari, Yann Le Cun and Sylvie Thiria*

Automata networks and artificial intelligence

7.1 Introduction

At the very beginnings of artificial intelligence (AI), one could find two main goals, which originally seemed to be considered as two aspects of only one problem: design 'intelligent' machines and understand human 'intelligence'. J. von Neumann [59] wanted a theory to establish the logical differences – if any – between 'natural' and 'artificial' automata. McCulloch and Pitts [54] had shown how formal neurons, similar to the nervous cells – the neurons – could be used to achieve the same computational performances as a Turing machine. The question of how to make them *learn* to compute was then raised by Rosenblatt [69] and Minsky and Papert [56] who tried to build a learning machine, the 'perceptron', based on such components. The issue in these early studies was very clear: would it be possible to build artificial systems capable of intelligence or, stated in other words, are people inherently smarter than machines? Perceptrons served as the prototype for all such systems, and a battlefield for controversy!

The first results could at first seem rather deceiving: while it is almost trivial to make neuron-based machines learn easy things (such as classifying objects into two categories), designing general-purpose learning machines appeared to be out of the range of the same techniques. As Rosenblatt states it: the perceptron 'does not generalize well to similar forms occurring in new positions in the retinal field, and its performance in detection experiments, where a familiar figure appears against an unfamiliar background, is apt to be weak'. These 'limitations' were made clear by Minsky and Papert [56] in their book which almost stopped any further research along these lines.

However, the ideas around the perceptron were used later on in the first developments of pattern recognition by Duda and Hart [17] and Nilsson [60]. But in the 1970s, AI mainly developed around tools based on logic [11]. Its achievements during this decade are amazing, and today we understand a lot more about our cognitive processes; but still people are much better than machines at certain tasks: understanding scenes or speech, playing chess or learning.

During the seventies the development of automata theory [16, 19, 26, 66] and more recently, work in statistical physics [45, 46, 61] has led to the reconsideration of these early models. At the same time, work on associative memories [29, 49, 57, 58] in connection with brain theory [18], has shown that the design of parallel architectures [35, 36, 43] might be based on the brain.

A new line of research based on automata networks – connectionism [20, 21, 22, 24, 25, 28, 38, 40] or parallel distributed processing [70] – is now developing, trying to push the limits imposed by the perceptron limitation theorem. More sophisticated machines have been designed – the Boltzmann Machine [1, 37, 74, 75], the Hierarchical Learning Machine [50, 51], layered feed-forward nets [63, 70] – that have already proved that general learning procedures could achieve what the perceptron could not. The applications of these new techniques cover many domains: vision [6, 8, 9, 10, 30, 55, 62, 73], cognitive science [3, 4, 53, 71, 72], language [68, 76, 79], learning [28, 41, 42, 51, 63, 70].

We will review in this chapter the different techniques used for learning by automata networks: the linear associator [48, 49], the perceptron [56, 69], threshold networks [45] and multi-layered networks [50, 51, 70, 76]. We will show how the dynamical behaviour of an automata network can be used to represent knowledge and realise associative memories, pattern recognition, automatic categorisation, structure extraction.

7.2   Adaptive machines

7.2.1   *The threshold automaton*

The first work on adaptive machines was based on the so-called *formal neuron* introduced by McCulloch and Pitts [54]. The formal neuron is also referred to as the *threshold automaton* [32] or linear threshold function (Fig. 7.1).

It is an automaton, that is a mapping $f: \mathscr{E}^n \to \mathscr{E}$, where $\mathscr{E}$ is the state space, (see [32, 67]). As such, it is interpreted as a cell which receives n inputs x_1, x_2, ..., x_n, taking their value in $\mathscr{E}$, and sends one output S. This output is computed as: $S = f(x_1, x_2, \ldots, x_n)$, i.e. by performing a weighted sum of the inputs, the output being 1 if this sum exceeds a threshold θ, and 0 otherwise:

$$S = \begin{cases} 1 & \text{if } \Sigma_j W_j x_j \geq \theta \\ 0 & \text{otherwise} \end{cases} \tag{7.1}$$

We will indifferently use this definition or the following, which can obviously be interchanged without any difficulty:

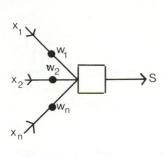
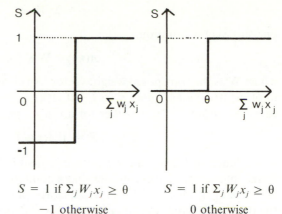

$$S = 1 \text{ if } \Sigma_j W_j x_j \geq \theta$$
$$-1 \text{ otherwise}$$

$$S = 1 \text{ if } \Sigma_j W_j x_j \geq \theta$$
$$0 \text{ otherwise}$$

Figure 7.1 Threshold automaton

$$S = \begin{cases} 1 & \text{if } \Sigma_j W_j x_j \geq \theta \\ -1 & \text{otherwise.} \end{cases} \tag{7.2}$$

Figure 7.1 shows an automaton (left) which receives inputs $x_1, \ldots, x_n$ along connections with weights $W_1, \ldots, W_n$, computes its state S and sends it as its output. S may be computed using state space $\mathscr{E} = \{-1, 1\}$ (middle) or $\mathscr{E} = \{0, 1\}$ (right).

The W_j are called the (connection) *weights* of the automaton, θ is called its *threshold*. In case (7.1), $\mathscr{E} = \{0, 1\}$, in case (7.2), $\mathscr{E} = \{-1, 1\}$.

7.2.2   *Learning the weights of a threshold automaton*

In this context, learning is just the problem of correctly adapting the weights of a linear threshold function. One of the first algorithms that could learn the weights was the *perceptron* learning rule proposed by Rosenblatt in 1957 [69]. However, a number of other learning rules are known [49] which are generally expressed in terms of a minimisation problem of a particular cost function. We will describe the Widrow–Hoff rule [81] also called Least Mean Square algorithm or *adaline* (adaptive linear element).

Assume that we want to teach a linear threshold automaton to classify a set of pattern vectors $\mathbf{X}_k, k = 1, \ldots, m$, elements of $\mathscr{E}^n$, into two classes C^+ and C^-. The class of pattern $\mathbf{X}_k$ (vectors and matrices will be printed in bold in the following, scalars in plain text) is labelled by a variable Y_k which can take one of the two values $+1$ for C^+ and -1 for C^-. The desired input-output mapping is thus given by the sequence of pairs $(\mathbf{X}_1, Y_1), (\mathbf{X}_2, Y_2), (\mathbf{X}_k, Y_k), \ldots (\mathbf{X}_m, Y_m)$.

The mapping of a linear threshold function – (7.1) and (7.2) – can be defined, using a *discriminant function* $A(\mathbf{x})$:

$$A(\mathbf{x}) = \Sigma_j W_j x_j \qquad (7.3)$$

or using vector notation:

$$A(\mathbf{x}) = \mathbf{W}'\mathbf{x} \qquad (7.4)$$

where $\mathbf{W}'$ is the transposed weight vector. An elementary classification can be implemented by assigning $\mathbf{x}$ to C^+ if $A(\mathbf{x}) \geq 0$, and to C^- otherwise. The effective output of the classifier for an input pattern $\mathbf{x}$ will then be:

$$S = \mathbf{1}[A(\mathbf{x})] \qquad (7.5)$$

where function $\mathbf{1}$ has value $+1$ if its argument is positive, and -1 otherwise. In the following, we will consider that $\mathbf{x}$ contains an additional coordinate x_0 whose value is always -1 and whose corresponding weight is interpreted as the threshold. Note that equation $A(\mathbf{x}) = 0$ defines a decision surface which is in our case a *hyperplane*, since $A(\mathbf{x})$ is linear.

Learning the weight vector can be viewed as minimising a cost function $C(\mathbf{W})$. Different choices can be made for C, each leading to a particular learning rule. One of the most widely used criteria is the mean squared error:

$$C(\mathbf{W}) = 1/(2m) \, \Sigma_k (\mathbf{W}' \cdot \mathbf{X}_k - Y_k)^2 \qquad (7.6)$$

where m is the number of patterns and factor 2 is introduced for notation convenience.

This criterion can be minimised by various techniques which will be extensively studied in the following sections. We will only consider here the simplest method, namely: the *gradient descent* procedure. The basic idea of gradient descent is the following. Take an arbitrary weight vector $\mathbf{W}_0$, and compute the gradient vector $\text{grad}_\mathbf{W}[C(\mathbf{W}_0)]$. The next value $\mathbf{W}_1$ is obtained by moving away from $\mathbf{W}_0$ along the direction of steepest descent which is just the negative of the gradient. This process is repeated until convergence, giving the following general formulation:

$$\mathbf{W}_{h+1} = \mathbf{W}_h - \lambda_h \, \text{grad}_\mathbf{W}[C(\mathbf{W}_h)] \qquad (7.7)$$

Here λ_h is a small positive factor that determines the step size towards the minimum. Since in our case $\text{grad}_\mathbf{W}[C(\mathbf{W})] = \Sigma_k [A(\mathbf{X}_k) - Y_k]\mathbf{X}_k$ we obtain the following learning rule:

$$\mathbf{W}_{h+1} = \mathbf{W}_h - \lambda_h \Sigma_k [A(\mathbf{X}_k) - Y_k]\mathbf{X}_k \qquad (7.8)$$

It can be shown that this process converges if λ_h is chosen 'sufficiently small' to avoid overshots [17, 49].

We can use the linearity property of the gradient operator to derive a simpler algorithm called *stochastic gradient*, which is essentially a noisy version of the original gradient descent. First, an infinite time sequence of pairs $(\mathbf{X}_k, Y_k)$ is built by repeatedly presenting the original sequence of patterns. At time h, a pattern pair $(\mathbf{X}^h, Y^h)$ is presented and an update of the weight vector $\mathbf{W}$ is performed according to the rule:

$$\mathbf{W}_{h+1} = \mathbf{W}_h - \lambda_h[A(\mathbf{X}^h) - Y^h]\mathbf{X}^h \qquad (7.9)$$

where h is now a time index. This is known as the *Widrow–Hoff rule*.

If a precise solution is not required, λ_h can be chosen constant, in which case the weight vector fluctuates around the optimal solution. These fluctuations increase with λ_h. Updating the network to add new information is easy since this adaptation rule proceeds continuously. In that case, the learning procedure minimises the *averaged* quadratic error.

A number of other criterion functions can be chosen that are suitable for gradient descent [17, 49, 77], including the so-called 'perceptron criterion':

$$C(\mathbf{W}) = -\frac{1}{2}\Sigma_k Y_k \cdot \mathbf{W}^t \cdot \mathbf{X}_k \qquad (7.10)$$

where the sum is taken over the *misclassified patterns* only. The resulting learning rule is simply:

$$\begin{cases} \mathbf{W}_{h+1} = \mathbf{W}_h + \lambda_h Y^h \cdot \mathbf{X}^h & \text{iff } \mathbf{X}^h \text{ is misclassified} \\ \mathbf{W}_{h+1} = \mathbf{W}_h & \text{otherwise} \end{cases} \qquad (7.11)$$

The parameter λ_h influences the length of the final weight vector but has no effect on its direction. The rule can be rewritten:

$$\mathbf{W}_{h+1} = \mathbf{W}_h - \lambda_h(S^h - Y^h)\mathbf{X}^h \qquad (7.12)$$

where S^h is the actual output when the input $\mathbf{X}^h$ is presented. This formula can be viewed as a *discrete* version of the Widrow–Hoff rule.

Unlike the Widrow–Hoff rule, the perceptron learning rule always finds a solution, provided that such a solution exists. Unfortunately when no exact solution exists, that is when the classes are not linearly separable, it does not even produce an approximate solution. The Widrow–Hoff rule is known to give a good approximate solution in both the separable and the nonseparable cases [17, 77]. The Widrow–Hoff rule has been extensively used in signal processing for implementing adaptive filters [44], and in pattern recognition for the synthesis of linear (or polynomial) classifiers.

In Section 7.3, a multidimensional version of the least square algorithm will be developed using the theoretical framework of matrix algebra. It will be applied to the problem of associative recall.

7.2.3 *The perceptron*

A *perceptron* $\mathbf{P}$ is a learning machine based on an automata network (Fig. 7.2) made of:

(i) A *retina* or set of cells where the input to the machine is displayed.

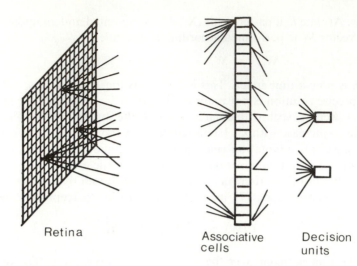

Figure 7.2 Perceptron. The figure shows a perceptron with a 2-dimensional retina, and 2 decision units.

Let $\mathscr{E}$ be the retina cells space, we will use $\mathscr{E} \subset \mathbb{R}$: for example, with $\mathscr{E} = \{-1, 1\}$, each cell can be 'on' ($x = 1$) or 'off' ($x = -1$); with $\mathscr{E} = \{0, 1, \ldots, k\}$ we can code the grey levels of each pixel in an image.

(ii) A layer of *associative* cells (or *feature extractors*): each cell is an automaton with transition function ψ_i defined on the retina: ψ_i: $\mathscr{E}^n \to \{-1, 1\}$, where n is the number of retina cells. Function ψ_i will be said to be of *connectivity* $k \leq n$ if ψ_i depends on only k variables in $\mathscr{E}^n$:

$$\exists i_1, \ldots, i_k \in \{1, \ldots, n\}: \forall x, y \in \mathscr{E}^n,$$
$$(x_j = y_j, \forall j \in \{i_1, \ldots, i_k\}) \Rightarrow \psi_i(x) = \psi_i(y)$$

(iii) A layer of *decision* units (usually one) which are threshold automata Ψ:

$$\Psi(x) = \begin{cases} 1 & \text{if } \Sigma_i W(\psi_i)\psi_i(\mathbf{x}) \geq \theta \\ -1 & \text{otherwise} \end{cases}$$

which we denote $\Psi = \mathbf{1}[\Sigma_i W(\psi_i)\psi_i - \theta]$, where the $W(\psi_i)$ are the *connection weights* of the associative cell i to the decision unit and θ the *threshold* of the decision unit.

Perceptrons are interesting because they can *learn* how to compute a predicate: let $\mathbf{P}$ be a perceptron structure (i.e. a set of retina cells, associative cells and decision units), then a predicate Ψ is *computable* by $\mathbf{P}$ if there exists a set of functions $\psi_1, \ldots, \psi_k$, weights $W^*(\psi_1), \ldots, W^*(\psi_k)$ and a threshold θ such that $\Psi = \mathbf{1}[\Sigma_i W^*(\psi_i)\psi_i - \theta]$. In this case, Minsky and Papert [56] have shown that weights $W(\psi_i)$ such that $\Psi = \mathbf{1}[\Sigma_i W(\psi_i)\psi_i - \theta]$ can be *learnt* by presenting examples. Let x be a family of 'examples', i.e. of vectors

x in $\mathcal{E}^n$; we will present a sequence of examples in the family in such a way that all examples are shown infinitely often. We then have the following theorem (this is the perceptron convergence theorem):

Theorem 1

Let Ψ be a predicate computable by a perceptron **P** and functions ψ_i. Then there exists a learning procedure which converges in a finite number of steps towards a set of weights $[W(\psi_i)]_i$ and threshold θ such that:

$$\Psi = \mathbf{1}[\Sigma_i W(\psi_i)\psi_i - \theta]$$

This learning procedure is given by:

(i) Choose an initial value: $[W_0(\psi_i)]_i$
(ii) At step k: choose an 'image' $\mathbf{x}(k)$ in $\mathcal{E}^n$ on the retina and compute $W_k(\psi_i)$, $\forall i$, by the following algorithm:

if $[(\Psi(\mathbf{x}(k)) = 1)$ **and** $(\Sigma_i W_k(\psi_i)\psi_i(\mathbf{x}(k)) < 0)]$ **then**

$$W_{k+1}(\psi_i): = W_k(\psi_i) + \psi_i(\mathbf{x}(k))$$

else

if $[(\Psi(\mathbf{x}(k)) = -1)$ **and** $(\Sigma_i W_k(\psi_i)\psi_i(\mathbf{x}(k)) \geq 0)]$**then**

$$W_{k+1}(\psi_i): = W_k(\psi_i) - \psi_i(\mathbf{x}(k));$$

$k: = k + 1;$

Remarks

(i) This algorithm just makes a change on $W_k(\psi_i)$ if pattern $\mathbf{x}(k)$ is misclassified. We have seen before, in Section 7.2.2, that this rule can be viewed as a gradient descent method (see (7.11)) and written as a discrete version of the Widrow–Hoff rule (7.12).

(ii) We have presented the learning algorithm in the case where $\theta = 0$. It is always possible to make this assumption by adding an auxiliary cell, which state is constantly set equal to -1 (Section 7.2.2). Its transition function is then $\psi = -1$ and Ψ can be rewritten as: $\Psi = \mathbf{1}[\Sigma_{i=1,\ldots,k} W(\psi_i)\psi_i + \theta\psi] = \mathbf{1}[\Sigma_{i=1,\ldots,k+1} W(\psi_i)\psi_i]$. The threshold θ is then learnt as an extra weight through the same algorithm. We will usually assume 0 threshold in the following, by eventually adding this auxiliary cell.

(iii) The limit weights $W_\infty(\psi_i) = \lim_{k\to\infty}[W_k(\psi_i)]$ may be different from the $W^*(\psi_i)$. For a computable predicate, there usually exists an infinite number of weight vectors to compute it. This is easily seen in Fig. 7.3: slightly moving around the line with equation $x + y - \frac{3}{2} = 0$ does not change the relative position of points $-1 - 1$, -11, $1 - 1$ and 11 with respect to the line and thus still provides a function that computes the '*and*' function.

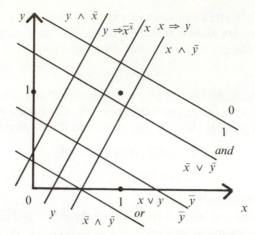

Figure 7.3 Representation of boolean functions as threshold functions

The figure shows the representation of the boolean functions in 2 variables, as threshold functions. For example, function *and* is represented by $\mathbf{1}[x + y - \frac{3}{2}]$: the figure shows the line with equation $x + y - \frac{3}{2} = 0$. It is easy to check that points 00, 01, and 10 are on the side of this line where $x + y - \frac{3}{2} < 0$, and thus the value of the threshold function – and of *and* – is 0. Point 11 is on the other side: the value of the threshold and *and* functions in 11 is 1.

(iv) The learning procedure is convergent for any choice of the initial condition $[W_0(\psi_i)]_i$ and image sequence $\mathbf{x}(k)$, which visits each example infinitely often.

(v) If $C^+ = \{\mathbf{x} \in \mathscr{E}^n: \Psi(\mathbf{x}) = 1\}$ and $C^- = \{\mathbf{x} \in \mathscr{E}^n: \Psi(\mathbf{x}) = -1\}$, then classes C^+ and C^- are said to be *separable* by Ψ with the perceptron **P**. The perceptron can thus be viewed as a machine for automatic classification: presented with 'examples' of class C^+ and examples of class C^-, it learns the weights $W_\infty(\psi_i)$, which will further allow it to *generalise* to any example of class C^+ or C^-. This means that, after learning, $\forall \mathbf{x} \in C^+$, $\Psi(\mathbf{x}) = 1$ and $\forall \mathbf{x} \in C^-$, $\Psi(\mathbf{x}) = -1$, even if such $\mathbf{x}$ has not been presented as an example in the learning process. It is this learning ability which is the main interest of perceptron-like machines, but the main problem is to decide whether a given predicate is computable by a given perceptron (i.e. the existence problem of preliminary weights $W^*(\psi_i)$).

(vi) Classification in more than two classes just requires the use of more than one decision unit.

We will say that a predicate $\Psi: \mathscr{E}^n \to \{-1, 1\}$ is of *order k* [56], if k is the smallest integer for which there exists a family of mappings $\psi_i: \mathscr{E}^n \to \{-1, 1\}$ such that $\forall i$, ψ_i is of connectivity $\leq k$, and $\Psi = \mathbf{1}[\Sigma_i W(\psi_i)\psi_i - \theta]$. Of course the order k of Ψ is always less than its connectivity.

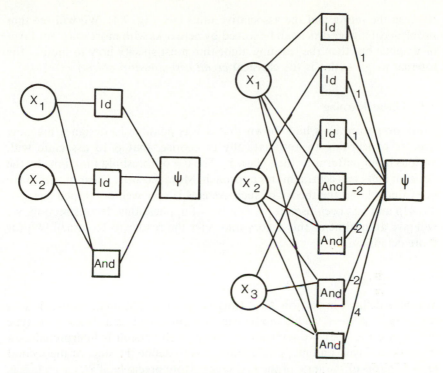

Figure 7.4 Computing the *xor* and the *n*-parity predicate with a perceptron

Figure 7.4 shows perceptrons that have been designed to compute the *xor* (left) and the 3-parity predicate (right). For example (in space $\mathscr{E} = \{0, 1\}$):

$$xor\ (x_1, x_2) = x_1 + x_2 - 2\ and\ (x_1, x_2)$$

Examples

(i) With $\mathscr{E} = \{0, 1\}$, 2 retina cells, 2 associative cells and one decision unit, we can realise all 16 boolean mappings in 2 variables by a perceptron. For 2 mappings: XOR and $\Leftrightarrow$, the associated perceptron will be of order 2, for the remaining 14 of order 1 only (see Fig. 7.3). This means that 14 boolean mappings in 2 variables can be realised by a threshold automaton [56].

For example: $\Psi(x, y) = x \vee y = or\ (x, y) = \mathbf{1}[x + y - \frac{1}{2}]$

(ii) The *n*-arity predicate: $\Psi(\mathbf{x}) = 1$ if card $\{i(1, \ldots, n): \mathbf{x}_i = 1\}$ is odd, is of order card $(\mathscr{E})$: for example, with $\mathscr{E} = \{0, 1\}$, $\Psi = $ XOR, which is of order 2.

It has been shown [56, 75] that the perceptron can only compute problems of first order, unless *ad hoc* connections specific to the problem are built between the retina and the associative units (see Fig. 7.4). We will see that problems of higher order can be solved by networks with more than one layer of weights, but then the learning algorithm must specify how to modify the internal weights (this is the classical *credit assignment problem*).

7.3   Linear learning

Work on perceptrons had shown that it was possible to design a machine capable of learning to automatically categorise, that is to associate with various input patterns $\mathbf{X}_i$ their class Y_i. The use of threshold functions by the decision units was central in this approach. More recently, *linear* models have been studied and the theory of such systems is now well developed (see [47, 48, 49]) and has been used for a number of applications. In this section, we will give an outline of this theory and refer the reader to Kohonen [49] for more details.

7.3.1   *The model*

We have defined in section A1.2 (this volume) an automata network as a mapping $F: \mathscr{E}^n \to \mathscr{E}^n$. This model can be represented as a 'black box' type model (see Fig. 7.5): we replicate $\mathscr{E}^n$ twice, each version is interpreted as a network or 'layer' and mapping F allows us to define the state of the second layer in terms of the state of the first layer. More precisely, if $\mathbf{x}(t)$, $t = 1, 2$, is the state vector of layer t, and j is a cell in the second layer, then j will have state:

$$x_j(2) = F_j[\mathbf{x}(1)]$$

Then the dynamics on F can be defined as usual, by feeding back the output of the second layer to the first one.

This representation allows us to extend the notion of automata network to mappings $F: \mathscr{E}^n \to \mathscr{E}^p$, with $n \neq p$ (see Fig. 7.5). The connections between the two layers defining F will be endowed with a *weight* $W_{ij} \in \mathbb{R}$. Let $\mathbf{W} = (W_{ij})$ be the weight matrix (if there is no connection between i and j, then by convention $W_{ij} = 0$) (Fig. 7.5)

We want to *memorise* into the network the *associations* between m input patterns $\mathbf{X}_i \in \mathscr{E}^n$, $i = 1, \ldots, m$, and m output patterns $_i \in \mathscr{E}^p$. Usually we will take $\mathscr{E} = \mathbb{R}$ or a finite space, $\{-1, 1\}$ for example. $\mathbf{X}$ will denote the $n \times m$ matrix with columns $\mathbf{X}_i$ and $\mathbf{Y}$ the $p \times m$ matrix with columns $\mathbf{Y}_i$: X_{ji} is thus the state of automaton j in pattern $\mathbf{X}_i$. In the case where $\mathbf{X}_i = \mathbf{Y}_i$, for all i, we will say that the network is performing *auto-association* (for example in pattern recognition, index searching), *hetero-association* otherwise (as in automatic classification).

The linear learning scheme assumes that these associations are realised by a *linear* automata network: when presented with an input pattern $\mathbf{x}$, the

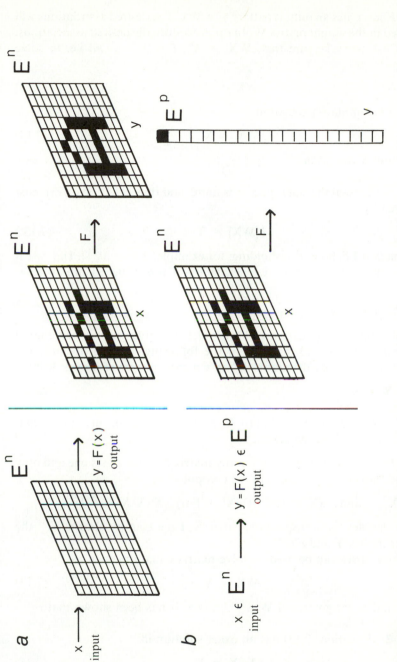

Figure 7.5 Automata network as a mapping (left) $F: \mathcal{E}^n \to \mathcal{E}^n$ (a) or $F: \mathcal{E}^n \to \mathcal{E}^p$ (b) and its equivalent (right) as a two-layered network. In the layered network representation, the input $\mathbf{x}$ to network F is the state of the input layer (represented here as a noisy 'A') and the output $\mathbf{y}$ is the state of the output layer (here $\mathbf{y}$ is a 'pure' A in case (a) and a code -1 for the first letter of the alphabet in case (b)).

network F computes an output pattern $\mathbf{y} = \mathbf{W}\mathbf{x}$. The desired associations will be encoded in the weight matrix $\mathbf{W}$. In order to store the desired associations, one thus has to make sure that: $\mathbf{W}\mathbf{X}_i = \mathbf{Y}_i$, $i = 1, \ldots, m$; i.e. to solve equation:

$$\mathbf{W}\mathbf{X} = \mathbf{Y} \qquad (7.13)$$

or in the case of auto-association:

$$\mathbf{W}\mathbf{X} = \mathbf{X} \qquad (7.14)$$

where $\mathbf{X}$ and $\mathbf{Y}$ are given.

Remark   In general the state space $\mathscr{E}$ is finite, and thus equation (7.13) must be replaced by:

$$\mathbf{1}[\mathbf{W}\mathbf{X}] = \mathbf{Y} \qquad (7.13')$$

where function $\mathbf{1}$ denotes thresholding: for example, if $\mathscr{E} = \{0, 1\}$, $\mathbf{1}[u] = 1$ if $u \geq 0$, 0 otherwise; if $\mathscr{E} = \{-1, +1\}$, $\mathbf{1}[u] = 1$ if $u \geq 0$, -1 otherwise.

7.3.2   *The generalised inverse matrix*

A formal solution to equation (7.13) or (7.14) is given by the generalised inverse matrix theory [13, 65, 78]. In the following, we will use only the pseudo-inverse [33]. By definition, $\mathbf{X}^+$ is a *pseudo-inverse* of matrix $\mathbf{X}$ iff:

(i) $\mathbf{X}\mathbf{X}^+\mathbf{X} = \mathbf{X}$

(ii) $\mathbf{X}^+\mathbf{X}\mathbf{X}^+ = \mathbf{X}^+$

(iii) $\mathbf{X}\mathbf{X}^+$ and $\mathbf{X}^+\mathbf{X}$ are hermitian (if $\mathbf{X}$ is a real matrix, this just means that $\mathbf{X}\mathbf{X}^+$ and $\mathbf{X}^+\mathbf{X}$ are symmetric)

It has been shown [33, 65] that for any matrix $\mathbf{X}$, there exists one and only one pseudo-inverse $\mathbf{X}^+$, which can be computed by:

$$\mathbf{X}^+ = \lim_{d \to 0}(\mathbf{X}'\mathbf{X} + d^2\mathbf{I})^{-1}\mathbf{X}' = \lim_{d \to 0}\mathbf{X}'(\mathbf{X}'\mathbf{X} + d^2\mathbf{I})^{-1}$$

where $\mathbf{X}'$ denotes the transpose of matrix $\mathbf{X}$, $\mathbf{I}$ the identity matrix, $\mathbf{Y}^{-1}$ the inverse of matrix $\mathbf{Y}$ and $d \in \mathbb{R}$.

Pseudo-inverses can be used to solve matrix equations such as:

$$\mathbf{W}\mathbf{X} = \mathbf{Y} \qquad (7.13)$$

where $\mathbf{X}$ and $\mathbf{Y}$ are given and $\mathbf{W}$ is unknown. It has been shown that:

Theorem 2   Equation (7.13) has an exact solution iff:

$$\mathbf{Y}\mathbf{X}^+\mathbf{X} = \mathbf{Y} \qquad (7.15)$$

and in this case, there exists an infinite number of solutions given by:

$$\mathbf{W} = \mathbf{YX}^+ + \mathbf{Z}(\mathbf{I} - \mathbf{XX}^+) \tag{7.16}$$

where $\mathbf{Z}$ is an arbitrary matrix with the same dimensions as $\mathbf{W}$.

Remarks

(i) Condition (7.15) is clearly necessary, since:

$$\mathbf{Y} = \mathbf{WX} \Rightarrow \mathbf{YX}^+\mathbf{X} = \mathbf{WXX}^+\mathbf{X} \Rightarrow \mathbf{YX}^+\mathbf{X} = \mathbf{WX} = \mathbf{Y}$$

(ii) If the columns $\mathbf{X}_i$ of $\mathbf{X}$ are linearly independent, then: $\mathbf{X}^+ = (\mathbf{X}'\mathbf{X})^{-1}\mathbf{X}' \Rightarrow$
$\mathbf{X}^+\mathbf{X} = \mathbf{I} \Rightarrow \mathbf{YX}^+\mathbf{X} = \mathbf{Y}$ and condition (7.15) is satisfied for all $\mathbf{Y}$. In this
case the solution $\mathbf{W} = \mathbf{YX}^+$ can be computed easily by computing only
$(\mathbf{X}'\mathbf{X})^{-1}\mathbf{X}'$.

Theorem 3 Among all solutions to equation (7.13), $\mathbf{W} = \mathbf{YX}^+$ is of
minimum quadratic norm.

Proof

$\|\mathbf{YX}^+ + \mathbf{Z}(\mathbf{I} - \mathbf{XX}^+)\|^2$
$\quad = \operatorname{tr}[\mathbf{YX}^+ + \mathbf{Z}(\mathbf{I} - \mathbf{XX}^+)]'[\mathbf{YX}^+ + \mathbf{Z}(\mathbf{I} - \mathbf{XX}^+)]$
$\quad = \|\mathbf{YX}^+\|^2 + \|\mathbf{Z}(\mathbf{I} - \mathbf{XX}^+)\|^2 + 2\operatorname{tr}[\mathbf{Z}'\mathbf{YX}^+] - 2\operatorname{tr}[(\mathbf{X}^+)'\mathbf{X}'\mathbf{Z}'\mathbf{YX}^+]$

where $\operatorname{tr}(\mathbf{A})$ is the trace of matrix $\mathbf{A}$.

By using the classical properties: $\operatorname{tr}(\mathbf{PQ}) = \operatorname{tr}(\mathbf{QP})$, $(\mathbf{X}^+)^+ = \mathbf{X}$ and
$\mathbf{XX}'(\mathbf{X}^+)' = \mathbf{X}$, we show that:

$$\operatorname{tr}[(\mathbf{X}^+)'\mathbf{X}'\mathbf{Z}'\mathbf{YX}^+] = \operatorname{tr}[\mathbf{Z}'\mathbf{YX}^+(\mathbf{X}^+)'\mathbf{X}'] = \operatorname{tr}[\mathbf{Z}'\mathbf{YX}^+]$$

and thus:

$$\|\mathbf{YX}^+ + \mathbf{Z}(\mathbf{I} - \mathbf{XX}^+)\|^2 = \|\mathbf{YX}^+\|^2 + \|\mathbf{Z}(\mathbf{I} - \mathbf{XX}^+)\|^2.$$

Matrix

$$\mathbf{W} = \mathbf{YX}^+ \tag{7.17}$$

is thus the solution to equation (7.13) of minimum quadratic norm. □

In the case of auto-association, the solution to equation (7.14) of
minimum quadratic norm is:

$$\mathbf{W} = \mathbf{XX}^+ \tag{7.17'}$$

If condition (7.15) is not satisfied, then equation (7.13) has no solution,
but it can be shown that matrices $\mathbf{W}$ for which the quadratic norm
$\|\mathbf{WX} - \mathbf{Y}\|$ is minimum are just those given by equation (7.16), and of
course, among those, matrix $\mathbf{W} = \mathbf{YX}^+$ is of minimum quadratic norm.
Hence, when (7.15) is not satisfied, (7.16) gives the general expression of
approximate solutions to (7.13) and (7.17) the best approximate solution.

Usually, the dimension of the state space $\mathscr{E}^n$ (i.e. the number of auto-mata in the network) is large – for example, in an image [49], $n = 3024$ (number of pixels) and $|\mathscr{E}| = 8$ (number of grey levels) – and thus presumably $m \ll n$ and the assumption of linear independence of the m patterns $\mathbf{X}_i$ is thus not very restrictive. In the case $m < n$, the operator $\mathbf{XX}^+$ is the orthogonal projection on the vector space $\mathscr{L}$ spanned by patterns $\mathbf{X}_1$, ..., $\mathbf{X}_m$ and $\mathbf{I} - \mathbf{XX}^+$ the orthogonal projection on $\mathscr{L}^\perp$. In the case of auto-association, $\mathbf{W}$ being just $\mathbf{XX}^+$, mapping F is then nothing more than the projection on $\mathscr{L}$.

7.3.3 *Computational techniques*

Exact methods to compute $\mathbf{X}^+$ have been derived. They are usually very time-consuming, since they involve computations of inverses of matrices. However, every computer maths library always includes an efficient program to compute the pseudo-inverse. We will present here various methods.

The first is a recursive algorithm, due to Greville [33], which does not need to invert matrices and may be implemented by *successively* presenting the examples $\mathbf{X}_1, \ldots, \mathbf{X}_m$: let us denote $\mathscr{X}_k = (\mathscr{X}_{k-1}\mathbf{X}_k)$ a matrix with k columns, where $\mathscr{X}_{k-1}$ is the submatrix of $\mathbf{X}$ formed with its $k-1$ first columns and $\mathbf{X}_k$ is the kth column of $\mathbf{X}$. We thus have: $\mathbf{X} = \mathscr{X}_m$. Greville's theorem states that $\mathbf{X}^+$ can be computed in m steps by:

$$\mathscr{X}_k^+ = \begin{bmatrix} \mathscr{X}_{k-1}^+(\mathbf{I} - \mathbf{X}_k\mathbf{p}_k') \\ \mathbf{p}_k' \end{bmatrix} \tag{7.18}$$

$$\text{and } \mathbf{p}_k = \begin{cases} \dfrac{(\mathbf{I} - \mathscr{X}_{k-1}\mathscr{X}_{k-1}^+)\mathbf{X}_k}{\|(\mathbf{I} - \mathscr{X}_{k-1}\mathscr{X}_{k-1}^+)\mathbf{X}_k\|^2} & \text{if the numerator is nonzero} \\[2ex] \dfrac{(\mathscr{X}_{k-1}^+)'\mathscr{X}_{k-1}^+\mathbf{X}_k}{1 + \|\mathscr{X}_{k-1}^+\mathbf{X}_k\|^2} & \text{otherwise} \end{cases}$$

As $\mathscr{X}_1$ is just the first column $\mathbf{X}_1$ of $\mathbf{X}$ we have:

$$\mathscr{X}_1^+ = \begin{cases} (\mathbf{X}_1'\mathbf{X}_1)^{-1}\mathbf{X}_1' & \text{if } \mathbf{X}_1 \text{ is nonzero} \\ 0 & \text{otherwise} \end{cases}$$

and of course $\mathbf{X}^+ = \mathscr{X}_m^+$.

The complexity of Greville's algorithm is of order n^2m^2, which may be a problem in practical situations. We will thus try to compute $\mathbf{W}$ directly instead of computing $\mathbf{X}^+$. A slight adaptation of Greville [33] allows us to recursively compute $\mathbf{W}$. By conserving the same notations for $\mathscr{X}_k$ and corresponding notations for $\mathscr{Y}_k$ and denoting $\mathbf{W}_k = \mathscr{Y}_k\mathscr{X}_k^+$, we have:

$$\mathbf{W}_k = \mathbf{W}_{k-1} - (\mathbf{W}_{k-1}\mathbf{X}_k - \mathbf{Y}_k)\mathbf{p}_k' \tag{7.19}$$

where $\mathbf{p}_k$ is the same vector as in Greville's theorem.

At iteration step k, $\mathbf{W}_k$ is the solution to equation $\mathbf{W}\mathcal{X}_k = \mathcal{Y}_k$ with minimum norm. This matrix thus encodes the associations between input patterns $\mathbf{X}_1, \ldots, \mathbf{X}_k$ and output patterns $\mathbf{Y}_1, \ldots, \mathbf{Y}_k$.

If the columns of $\mathbf{X}$ are independent, the computation of $\mathbf{W}$ becomes simpler. We then have:

$$\mathbf{W}_k = \begin{cases} \mathbf{W}_{k-1} - (\mathbf{W}_{k-1}\mathbf{X}_k - \mathbf{Y}_k)[\chi_k'/\|\chi_k\|^2] & \text{if } \chi_k \neq 0 \\ \mathbf{W}_{k-1} & \text{otherwise} \end{cases} \quad (7.20)$$

with:

$$\begin{array}{ll} \chi_k = \mathbf{S}_{k-1}\mathbf{X}_k & \chi_1 = \mathbf{X}_1 \\ \mathbf{S}_k = \mathbf{S}_{k-1} - \chi_k\chi_k'/\|\chi_k\|^2 & \mathbf{S}_0 = \mathbf{I} \end{array} \quad (7.20')$$

The χ_k can also be computed through the Gram–Schmidt orthogonalisation process:

$$\begin{aligned} \chi_1 &= \mathbf{X}_1 \\ \chi_k &= \mathbf{X}_k - \sum_{i=1}^{k-1} \langle \mathbf{X}_k, \chi_i \rangle \cdot \chi_i/\|\chi_i\|^2 \quad k = 2, \ldots, m \end{aligned} \quad (7.20'')$$

where the summation runs on all nonzero χ_i. This result is similar to Pyle [64]. More details can be found in Kohonen [49].

7.3.4 *Approximated computation*

The preceding algorithms may imply a huge computational burden if the dimension n of the patterns is large. Moreover, simulations show that the linear learning scheme is very fault-tolerant (see below): even approximate values of matrix $\mathbf{W}$ would do. Therefore, approximation methods are used to estimate $\mathbf{W}$: instead of computing the exact matrix $\mathbf{Y}\mathbf{X}^+$, we compute a matrix $\mathbf{W}$ which minimises the error $\mathbf{E}$ between the output value $\mathbf{W}\mathbf{X}$ and the desired value $\mathbf{Y}$. Of course many error functions can be used, the most classical being the quadratic norm which requires the minimisation of:

$$\mathbf{E} = \sum_{k=1}^{m} \sum_{i=1}^{p} \left(\sum_{j=1}^{n} W_{ij}X_{jk} - Y_{ik} \right)^2 \quad (7.21)$$

This is a classical minimisation problem, which can be solved by using gradient methods: $\text{grad}_w \mathbf{E} = 2(\mathbf{W}\mathbf{X} - \mathbf{Y})\mathbf{X}'$ is the gradient of $\mathbf{E}$ with respect to $\mathbf{W}$. We have presented various methods in Section 7.2.2: for example, the gradient method with increment (λ_k) is given by:

$$\begin{cases} \mathbf{W}_0 \text{ arbitrary} \\ \mathbf{W}_k = \mathbf{W}_{k-1} - \lambda_k(\mathbf{W}_{k-1}\mathbf{X} - \mathbf{Y})\mathbf{X}' \end{cases} \quad (7.22)$$

This procedure has been shown to converge provided the λ_k are chosen such that:

$$\sum_{k=1}^{\infty} \lambda_k = +\infty \text{ and } \sum_{k=1}^{\infty} \lambda_k^2 < \infty$$

This is satisfied in particular for $\lambda_k = \lambda_1/k$, but convergence may be very slow.

One weakness, for our purpose, of these gradient methods is that they require the full knowledge of matrices $\mathbf{X}$ and $\mathbf{Y}$. We would prefer instead methods that allow the 'knowledge' of the network to gradually evolve when new data are presented, without of course destroying previous memories. In other words, we want to design a method that would allow us to compute $\mathbf{W}_k$ in terms of $\mathbf{W}_{k-1}$, by making the correction needed to take into account the observed error when retrieving the new pattern $\mathbf{X}_k$ by $\mathbf{W}_{k-1}$. This can be done by using a correction term proportional to $(\mathbf{W}_{k-1}\mathbf{X}_k - \mathbf{Y}_k)$, which is the discrepancy between the desired output and the output computed with $\mathbf{W}_{k-1}$. This gives the following algorithm:

$$\mathbf{W}_k = \mathbf{W}_{k-1} - (\mathbf{W}_{k-1}\mathbf{X}_k - \mathbf{Y}_k)\mathbf{c}_k^{\prime} \tag{7.23}$$

where $\mathbf{c}_k^{\prime}$ is a vector that will allow us to adapt the weights to the new incoming pattern $\mathbf{X}_k$. If the columns $\mathbf{X}_k$ are linearly independent, then one possible choice is: $\mathbf{c}_k = \mathbf{p}_k$, and then, at each iteration step, matrix $\mathbf{W}_k$ is optimal for patterns $\mathbf{X}_1, \ldots, \mathbf{X}_k$ and adding one more pattern just requires us to iterate formula (7.23). This is just the case discussed in (7.19).

In the general case, going in the direction computed with $\mathbf{c}_k = \lambda \mathbf{X}_k$ decreases the error, since $(\mathbf{W}_{k-1}\mathbf{X}_k - \mathbf{Y}_k)\mathbf{X}_k^{\prime}$ is the gradient of the error $\mathbf{E}_k$ on the kth pattern $\mathbf{X}_k$.

There also exist other methods, which do not require the optimality of matrix $\mathbf{W}_k$ at each iteration step: in 'stochastic' gradient methods, each step minimises the error on one pattern only. One of the most famous of these stochastic methods is the Widrow–Hoff procedure [81]: patterns $\mathbf{X}_1, \ldots, \mathbf{X}_m$ are introduced sequentially and repeatedly, until convergence. To simplify notation, we will denote this sequence $(\mathbf{X}_1, \mathbf{Y}_1), (\mathbf{X}_2, \mathbf{Y}_2), \ldots, (\mathbf{X}_m, \mathbf{Y}_m), (\mathbf{X}_1, \mathbf{Y}_1), \ldots$ by $(\mathbf{X}^1, \mathbf{Y}^1), (\mathbf{X}^2, \mathbf{Y}^2), \ldots, (\mathbf{X}^m, \mathbf{Y}^m), (\mathbf{X}^{m+1}, \mathbf{Y}^{m+1}), \ldots$ At each step, a new pattern is presented and the error on this pattern is minimised: it is necessary to go in the opposite direction from the gradient of the error $\mathbf{E}_{i+1}$ on pattern $\mathbf{X}^{i+1}$. This gives the following rule:

$$\begin{cases} \mathbf{W}^0 \text{ arbitrary} \\ \mathbf{W}^{i+1} = \mathbf{W}^i - \lambda_i(\mathbf{W}^i\mathbf{X}^{i+1} - \mathbf{Y}^{i+1})[\mathbf{X}^{i+1}]^{\prime} \end{cases} \tag{7.24}$$

There is no existing proof of the convergence of this algorithm. However, in practice, it has been observed that with $\lambda_i = \lambda_1/i$, the algorithm usually converges to a solution. The problem of adding one pattern to the memory is quite easy, it is sufficient to start with $\mathbf{W}^0 = \mathbf{W}$ and iterate with the new pattern until convergence. As the starting matrix is probably close to the solution, the computing time will not be long before convergence.

7.3.5   *Other models*

Constrained gradients

It is sometimes useful to impose on the connection matrix some constraints that reflect properties of the information flows in the network. For example, the connection weight of an element with itself, W_{ii}, could be bounded, or even cancelled, so that its influence would not overkill the others. In this case, matrix **W** is constrained to have a bounded – or zero – diagonal and can be computed by gradient projection methods. In the case of $W_{ii} = 0$, there exist solutions to equation (7.13) of the form: $\mathbf{W} = \mathbf{YX}^+ + \mathbf{Z}(\mathbf{I} - \mathbf{XX}^+)$.

Hebb's law

In 1949, Hebb [34] proposed his 'reinforcement law' based on biological evidence: any two neurons firing at the same time, when 'shown' a given pattern, will have their connection strength reinforced. This is also related to the theory of learning by evolution of the synapse structure (see Changeux [14]). Stated in informal fashion by Hebb, this rule has been repeatedly used to compute weights W_{ij}, through algorithms such as:

$$W_{ij} = W_{ij} + X_{ik}X_{jk} \tag{7.25}$$

if pattern $\mathbf{X}_k$ is shown to the network: $X_{ik} = 1$ if automaton i is on in pattern $\mathbf{X}_k$, -1 otherwise. In the case of auto-association between patterns $\mathbf{X}_1$, $\mathbf{X}_2$, ..., $\mathbf{X}_m$, algorithm (7.25) is successively applied to all patterns and the corresponding weights added. Matrix **W** is thus given by:

$$\mathbf{W} = \mathbf{XX}^t \tag{7.26}$$

In the case of hetero-association between patterns $(\mathbf{X}_1, \mathbf{Y}_1)$, $(\mathbf{X}_2, \mathbf{Y}_2)$, ..., $(\mathbf{X}_m, \mathbf{Y}_m)$, this rule can be extended: $W_{ij} = W_{ij} + Y_{ik}X_{jk}$ and $\mathbf{W} = \mathbf{YX}^t$.

Note that, when patterns $\mathbf{X}_k$ are linearly independent and orthonormal, that is $\langle \mathbf{X}_i, \mathbf{X}_j \rangle = 0$, $\forall i \neq j$, and $\|\mathbf{X}_k\| = 1$, then $\mathbf{X}^+ = \mathbf{X}^t$ and $\mathbf{W} = \mathbf{YX}^t$ is the solution of minimum norm of equation (7.13). Hence Hebb's rule, based on biological evidence, can be considered as an approximated version of the theoretical results in Section 7.3.2.

7.4   **Retrieval**

7.4.1   *Introduction*

After the learning session, during which the network connection matrix **W** has been computed, we can use the network to perform recognition tasks. The main interest of these learning techniques is that they lead to efficient retrieval even if the data are noisy or partial. This will thus allow us to:

(i) classify patterns close to memorised items,
(ii) retrieve memorised items from noisy or incomplete patterns.

That is to say, we can implement a function of *associative memory*.

Various procedures have been designed for retrieval: remember that our learning procedure was based on a linear network, but, as can be expected, nonlinear retrieval rules may be used for improved performance.

7.4.2   *Linear retrieval*

We will now discuss, in the case of a linear retrieval process, the properties of resistance to noise depending on which model of noise is applied to the inputs: we will derive the 'best' choice of matrix $\mathbf{W}$ for various models of noise.

We have seen that the output of a linear network, with connection matrix $\mathbf{W}$, when the input is some vector $\mathbf{x}$, is $\mathbf{y} = \mathbf{Wx}$. If $\mathbf{x} = \mathbf{X}_k$, then $\mathbf{y}$ will be exactly $\mathbf{Y}_k$. If $\mathbf{x}$ is only close to some $\mathbf{X}_k$, then $\mathbf{y}$ will be close to $\mathbf{Y}_k$. Suppose

$$\mathbf{W} = \mathbf{YX}^+$$

Then

$$\mathbf{y} = \mathbf{YX}^+\mathbf{x} = \mathbf{YX}^+[(\mathbf{XX}^+)\mathbf{x}]$$

(because $\mathbf{X}^+\mathbf{XX}^+ = \mathbf{X}^+$, by the definition of the pseudo-inverse)

$$\Rightarrow \qquad\qquad \mathbf{y} = \mathbf{YX}^+\mathbf{x}_{\mathscr{L}}$$

where $\mathbf{x}_{\mathscr{L}}$ denotes the projection of $\mathbf{x}$ onto the vector space $\mathscr{L}$ spanned by the $\mathbf{X}_k$ (see Section 7.3.1):

$$\mathbf{x}_{\mathscr{L}} = \sum_{i=1}^{m} \alpha_i \mathbf{X}_i$$

$$\Rightarrow \qquad \mathbf{y} = \mathbf{YX}^+ \sum_{i=1}^{m} \alpha_i \mathbf{X}_i = \sum_{i=1}^{m} \alpha_i \mathbf{YX}^+ \mathbf{X}_i = \sum_{i=1}^{m} \alpha_i \mathbf{Y}_i$$

But if $\mathbf{x}$ is close to $\mathbf{X}_k$, then α_k is large compared to the other α_i, and thus $\mathbf{y}$ is close to $\mathbf{Y}_k$.

Simulations (see Kohonen [49]) show that results are better if the memorised patterns are close to being mutually orthogonal, thus any preprocessing which tends to orthogonalise the patterns would lead to increased efficiency.

The signal-to-noise ratio is improved after retrieval.

Proof   Suppose $\mathbf{x} = \mathbf{X}_k + \mathbf{v}$, where $\mathbf{v}$ is some noise vector. Then:

$$\mathbf{y} = \mathbf{Y}\mathbf{X}^+\mathbf{x} = \mathbf{Y}\mathbf{X}^+[(\mathbf{X}\mathbf{X}^+)(\mathbf{X}_k + \mathbf{v})] \quad (\text{because } \mathbf{X}^+\mathbf{X}\mathbf{X}^+ = \mathbf{X}^+)$$

$$\Rightarrow \quad \mathbf{y} = \mathbf{Y}\mathbf{X}^+[\mathbf{x}_{\mathscr{L}} + \mathbf{v}_1]$$

where $\mathbf{v}_1$ denotes the orthogonal projection of $\mathbf{v}$ on $\mathscr{L}$. $\qquad \square$

It has been shown [49], for auto-association, that if $\mathbf{v}$ is of constant length and has a direction uniformly distributed in $\mathbb{R}^n$ then the variance:

$$\text{var}(\|\mathbf{W}\mathbf{x} - \mathbf{X}_k\|) = (m/n)\|\mathbf{x} - \mathbf{X}_k\|$$

Hence, if $m < n$, the noise is damped in the orthogonal projection. In the case of hetero-association, performances are usually quite good, but the output noise-to-signal ratio may eventually become large even for small input noise-to-signal ratio, when m/n increases and the determinant of $\mathbf{X}\mathbf{X}'$ goes to 0 [78].

More generally, suppose $\mathbf{x} = \mathbf{X}_k + \mathbf{v}$, then $\mathbf{y} = \mathbf{W}\mathbf{x} = \mathbf{W}(\mathbf{X}_k + \mathbf{v}) = \mathbf{W}\mathbf{X}_k + \mathbf{W}\mathbf{v}$. Hence, to improve the efficiency in terms of signal-to-noise ratio, we can try to find the matrix $\mathbf{W}$ which minimises the dispersion of the remaining noise at the output: $\mathbf{W}\mathbf{v}$.

If $\mathbf{W}$ is of the form of equation (7.16) and $\mathbf{v}$ is of zero mean, then the dispersion of the remaining noise at the output is:

$$\text{var}(\mathbf{W}\mathbf{v}) = E[\mathbf{W}\mathbf{v}(\mathbf{W}\mathbf{v})'] = E[\mathbf{W}\mathbf{v}\mathbf{v}'\mathbf{W}'] = \mathbf{W}E[\mathbf{v}\mathbf{v}']\mathbf{W}' = \mathbf{W}\Sigma\mathbf{W}'$$

where Σ is the dispersion matrix of noise $\mathbf{v}$.

$$\Rightarrow \quad \text{var}(\mathbf{W}\mathbf{v}) = [\mathbf{M}_0 + \mathbf{Z}\mathbf{M}_1]\Sigma[\mathbf{M}_0 + \mathbf{Z}\mathbf{M}_1]' \text{ with } \mathbf{M}_0 = \mathbf{Y}\mathbf{X}^+$$

$$= [\mathbf{M}_0 + \mathbf{Z}\mathbf{M}_1]\Sigma[\mathbf{M}_0' + \mathbf{M}_1'\mathbf{Z}'] \text{ and } \mathbf{M}_1 = \mathbf{I} - \mathbf{X}\mathbf{X}^+$$

$$= \mathbf{M}_0\Sigma\mathbf{M}_0' + \mathbf{M}_0\Sigma\mathbf{M}_1\mathbf{Z}' + \mathbf{Z}\mathbf{M}_1\Sigma\mathbf{M}_0' + \mathbf{Z}\mathbf{M}_1\Sigma\mathbf{M}_1\mathbf{Z}'$$

$$= \mathbf{M}_0\Sigma\mathbf{M}_0' + \mathbf{A}\mathbf{Z}' + \mathbf{Z}\mathbf{A}' + \mathbf{Z}\mathbf{B}\mathbf{Z}'$$

with evident notations for $\mathbf{A}$ and $\mathbf{B}$. Note that $\mathbf{M}_1$ and $\mathbf{B}$ are symmetric. Let V_{ij} be the coefficient in line i and column j of matrix var $(\mathbf{W}\mathbf{v})$. We have:

$$V_{ij} = (\mathbf{M}_0\Sigma\mathbf{M}_0')_{ij} + \sum_{k=1}^{n} A_{ik}(\mathbf{Z}')_{kj} + \sum_{k=1}^{n} Z_{ik}A_{jk}$$

$$+ \sum_{k=1}^{n} Z_{ik} \sum_{\ell=1}^{n} B_{k\ell}(\mathbf{Z}')_{\ell j}$$

$$= (\mathbf{M}_0\Sigma\mathbf{M}_0')_{ij} + \sum_{k=1}^{n} A_{ik}Z_{jk}$$

$$+ \sum_{k=1}^{n} A_{jk}Z_{ik} + \sum_{k,\ell=1}^{n} B_{k\ell}Z_{ik}Z_{j\ell}$$

To minimise var$(\mathbf{W}\mathbf{v})$, it is thus sufficient to choose $\mathbf{Z}$ such that:

$$\partial V_{ij}/\partial Z_{\alpha\beta} = 0 \quad \forall\alpha, \beta$$

with:

$$\partial V_{ij}/\partial Z_{\alpha\beta} = 0 \quad \forall \alpha \neq j, i$$

$$\partial V_{ij}/\partial Z_{i\beta} = \delta_{ij}A_{i\beta} + A_{j\beta} + . \sum_{\ell=1}^{n} B_{\beta\ell}Z_{j\ell} + \delta_{ij}\sum_{k=1}^{n} B_{k\beta}Z_{ik}$$

$$\partial V_{ij}/\partial Z_{j\beta} = A_{i\beta} + \delta_{ij}A_{j\beta} + \delta_{ij}\sum_{\ell=1}^{n} B_{\beta\ell}Z_{j\ell} + \sum_{k=1}^{n} B_{k\beta}Z_{ik}$$

where δ_{ij} is 1 iff $i = j$, 0 otherwise. Thus $\mathbf{Z}$ satisfies the equation:

$$\mathbf{ZB} = -\mathbf{A} \text{ or } \mathbf{ZM_1 \Sigma M_1} = -\mathbf{M_0 \Sigma M_1}$$

which is of the same kind as equation (7.13). Thus there exists a matrix $\mathbf{Z}$ of minimum quadratic norm which minimises the dispersion of the noise at the output; it is given by: $\mathbf{Z} = -\mathbf{AB^+}$.

If noise $\mathbf{v}$ has a diagonal dispersion matrix $\Sigma = \sigma^2\mathbf{I}$, then $\mathbf{Z}$ is such that:

$$\mathbf{ZM_1\Sigma M_1} + \mathbf{M_0\Sigma M_1} = 0 \Rightarrow \mathbf{Z}[\mathbf{I} - \mathbf{XX^+}][\mathbf{I} - \mathbf{XX^+}]\sigma^2 + \mathbf{YX^+}[\mathbf{I} - \mathbf{XX^+}]\sigma^2 = 0$$

$$\Rightarrow \quad \mathbf{Z}[\mathbf{I} - \mathbf{XX^+}] = 0 \text{ (if } \sigma^2 \neq 0) \Rightarrow \mathbf{Z} = 0 \text{ or } \mathbf{I} - \mathbf{XX^+} = 0$$

Hence, for a noise $\mathbf{v}$ with 0 mean and dispersion $\sigma^2\mathbf{I}$, the solution to (7.13) $\mathbf{W} = \mathbf{YX^+}$ (with $\mathbf{Z} = 0$) minimises the remaining noise at the output. $\quad\square$

In many practical situations, the state space $\mathscr{E}$ is not $\mathbb{R}$, but a discrete space, for example $|\mathscr{E}| = r$ for an image in $\mathbf{r}$ grey levels. Then, the input patterns $\mathbf{X}_k$ are in this discrete state space, but $\mathbf{W}$ has its elements in $\mathbb{R}$ and thus the output $\mathbf{y}$ is in $\mathbb{R}^p$. The output must then be thresholded so as to be interpreted in the same terms as the inputs (e.g. image). Usually, performances will be improved, in the case of auto-association, if the thresholded output is fedback to the network and the whole process of projection and thresholding is reiterated (Fig. 7.6): this can be seen when the linear operator is used. With input $\mathbf{x}$, the projection $\mathbf{XX^+}$ builds vector $\mathbf{x}_{\mathscr{L}}$ which may lie in between the memorised items $\mathbf{X}_i$, very close – but not identical – to some $\mathbf{X}_k$. As further projections will keep $\mathbf{x}_{\mathscr{L}}$ fixed, there will be no way to move $\mathbf{x}_{\mathscr{L}}$ towards $\mathbf{X}_k$: repeated thresholding, by taking $\mathbf{x}_{\mathscr{L}}$ out of $\mathscr{L}$ will allow further approximations to take place, thus improving performances. This leads us to retrieval processes which are in fact far from linear and there does not exist any more theoretical reason why matrix $\mathbf{W} = \mathbf{YX^+}$ should be optimal for these retrieval processes.

Figure 7.6 shows a network, with connection matrix $\mathbf{W}$. The retrieval process consists in sending the input, $\mathbf{X}_k$ in the figure, through the network that computes $\mathbf{WX}_k$, then thresholding this result, component by component, then feeding back the thresholded value into the network. The process can be reiterated until a fixed point is reached, or stopped after a certain number of iterations.

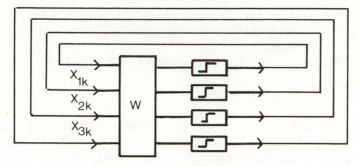

Figure 7.6 An iterative network

7.4.3 *Threshold networks*

The model of threshold networks has been largely studied [27, 31, 45, 52, 61, 80] almost independently from the work presented so far in the previous sections. However, when those networks are proposed to implement associative memories [27, 45], it is easy to see, as we mentioned before (Section 7.3.5), that they are very close to the linear model, with thresholded output: when patterns $\mathbf{X}_k$ are linearly independent and orthonormal, then the optimal associative memory, that is the solution of minimum quadratic norm of equation (7.13) $\mathbf{WX} = \mathbf{Y}$ is given by (7.26): $\mathbf{W} = \mathbf{XX}^t$. This rule, inspired from Hebb's reinforcement law [34], was first introduced by Little [52] and further elaborated by Hopfield [45] who proposed to use an *energy* function.

One very distinctive feature of the approach developed below is the use of *iterations*: in the linear retrieval model, a pattern $\mathbf{x}$ is projected (in the case of auto-association) onto $\mathscr{L}$, and this gives the final output. We will present here models where the output is the result of an iteration process run on the network. Various kinds of iteration modes have been defined on networks [67] which might eventually lead to different results.

In this section, we will fully develop this model and show how tools introduced in chapter 1 and [32, 67] can be used to give theoretical results on the retrieval performances. The material presented here can be found in more detail in [23, 31].

Using the definition of a threshold automaton (7.1), we have:

Definition We will say that $F: \mathscr{E}^n \to \mathscr{E}^n$ is a *threshold network* iff all its components $(f_1, \ldots, f_n)$ are threshold automata (see (7.1) and (7.2)):

$$\forall i(1, \ldots, n), \forall \mathbf{x} \in \mathscr{E}^n, f_i(\mathbf{x}) = \begin{cases} 1 & \text{if } \Sigma_j \, W_{ij} x_j - \theta_i \geq 0 \\ -1 & \text{otherwise} \end{cases}$$

where $\mathbf{W} = (W_{ij})$ is a real $n \times n$ matrix and $\boldsymbol{\theta} = (\theta_i)$ is a real n-vector, denoting, as usual, the weights and thresholds of the network.

In the following, we will use, unless specifically mentioned, the case where $\mathcal{E} = \{0, 1\}$. We have seen before that this case could easily be rewritten into the $\mathcal{E} = \{-1, 1\}$ case.

f_i is called a *strict* threshold automaton (see Chapter 4) iff:

$$f_i(\mathbf{x}) = \begin{cases} 1 & \text{if } \Sigma_j\, W_{ij}x_j - \theta_i > 0 \\ -1 & \text{if } \Sigma_j\, W_{ij}x_j - \theta_i < 0 \end{cases}$$

It is easy to show that any threshold automaton is equivalent to a strict threshold automaton [31] by changing the thresholds. We will thus, in the following, make the assumption of strict threshold, which is necessary in some proofs.

It has been shown [66, 67] that different modes of iteration could be defined on a network. We will use the following notations:

Notation   Let $F: \mathcal{E}^n \rightarrow \mathcal{E}^n$ be an automata network of n elements and $\mathbf{x}^0 \in \mathcal{E}^n$. The different *iteration modes* on F with initial condition $\mathbf{x}^0$ are defined by $\mathbf{x}^0 = \mathbf{x}(0)$ and $\mathbf{x}(t) \in \mathcal{E}^n$, the state of F at time t, where $\mathbf{x}(t)$ is:

 (i) *Parallel iteration:* $\forall t \geq 0,\ \mathbf{x}(t + 1) = F[\mathbf{x}(t)]$
 (ii) *Sequential iteration* associated to the permutation π of $\{1, \ldots, n\}$:

$$\forall t \geq 0,\ \mathbf{x}(t + 1) = F_\pi[\mathbf{x}(t)]$$

 Elements i of the network change state one at a time in the order prescribed by π.
 (iii) *Block-sequential iteration* associated to a partition $(I_k)_{k=1,\ldots,p}$ of $\{1, \ldots, n\}$, ordered (which means that: $\forall \alpha \in I_i, \forall \beta \in I_j, i < j \Rightarrow \alpha < \beta$) as:

$$\mathbf{x}(t + 1) = F_{(I_k)}[\mathbf{x}(t)]$$

 Elements i in I_k change state in parallel, and the blocks I_k are iterated serially one after the other.
 (iv) *Random iteration*, with visiting strategy $(i_t)_{t \geq 0}$ (where $(i_t)_{t \geq 0}$ is a random process with values in $\{1, \ldots, n\}$):

$$\forall t \geq 0,\ \mathbf{x}(t + 1) = F_{(i_t)}[\mathbf{x}(t)].$$

 Elements i of the network change state one at a time in the order prescribed by i_t.

Definition   Let $F: \mathcal{E}^n \rightarrow \mathcal{E}^n$ be a threshold network. We define the *energy* of the network for the different iteration modes as follows:

 (i) Sequential, block-sequential or random iterations: $\forall \mathbf{x} \in \mathcal{E}^n$,

$$E(\mathbf{x}) = -\frac{1}{2} \sum_{i=1}^{n} x_i \sum_{j \neq i} W_{ij}x_j + \sum_{i=1}^{n} (\theta_i - W_{ii})x_i$$

 (ii) Parallel iteration: $\forall \mathbf{x} \in \mathcal{E}^n,\ E[\mathbf{x}(t)] = \xi[\mathbf{x}(t), \mathbf{x}(t - 1)]$

with:
$$\xi[\mathbf{u}, \mathbf{v}] = \sum_{i=1}^{n} u_i \sum_{j=1}^{n} W_{ij} v_j + \sum_{i=1}^{n} \theta_i [u_i + v_i]$$

Remark   This energy coincides with the energy of a spin glass with field θ, as defined for example in [45], only in the case of sequential, block-sequential and random iterations. The energy for the parallel iteration does not have any physical meaning in this context (see also [32]).

The importance of the concept of energy is due to the following theorem (see [32] and chapter 4 for proofs):

Theorem 4   Let $F: \mathscr{E}^n \to \mathscr{E}^n$ be a threshold network.

Then: $\forall t \geq 0, \mathbf{x}(t + 1) \neq \mathbf{x}(t) \Rightarrow E[\mathbf{x}(t + 1)] < E[\mathbf{x}(t)]$

 (i) for any sequential iteration, where $\mathbf{W}$ is a symmetric matrix, with nonnegative diagonal elements.
(ii) for any block-sequential iteration associated to the ordered partition (I_k), where $\mathbf{W}$ is a symmetric matrix with its blocks $\mathbf{W}_k = (W_{ij})_{i,j \in I_k}$ nonnegative definite on the set $\{-1, 0, 1\}$, for all k.
(iii) for the parallel iteration, where $\mathbf{W}$ is a symmetric matrix.

(In each case, E is the energy defined for that iteration mode.)

This theorem thus shows that the energy is a *Lyapunov function* of the network for the different iteration modes, which leads to:

Corollary   Under the conditions of the previous theorem:

(i) All sequential, random and block-sequential iterations on threshold networks only have *fixed points*.

Furthermore, the *transient length* of any trajectory is bounded below:

$$\forall \mathbf{x}^0 \in \mathscr{E}^n, \; t(\mathbf{x}^0) \geq [1/\varepsilon] \left[\frac{1}{2} \sum_{i,j=1}^{n} |W_{ij}| + \sum_{i=1}^{n} |\theta_i| \right]$$

with $\varepsilon = \min\{|E[\mathbf{x}(t)] - E[\mathbf{x}(t + 1)]|/\mathbf{x}^0 \in \mathscr{E}^n, \mathbf{x}(t + 1) \neq \mathbf{x}(t)\}$
Moreover, if $\mathbf{W}$ is an integer matrix ($W_{ij} \in \mathbb{Z}$):

$$t(\mathbf{x}^0) \leq \sum_{i,j} |W_{ij}| + 2\sum_{i} |\theta_i|$$

and if $\mathbf{W}$ has its elements in $\{-1, 0, 1\}$ ($W_{ij} \in \{-1, 0, 1\}$):

$$t(\mathbf{x}^0) \leq 3nv$$

where $\forall i$, card $\{j \in \{1, \ldots, n\}: W_{ij} \neq 0\} \leq v$ $(v \leq n)$
(ii) The parallel iteration on a threshold network only has limit cycles of period 1 or 2. Furthermore, the transient length of any trajectory is bounded below:

$$\forall \mathbf{x}^0 \in \mathscr{E}^n, \ t(\mathbf{x}^0) \le [1/\varepsilon]\left[\frac{1}{2}\sum_{i,j}|W_{ij}| + \sum_i |\theta_i|\right]$$

with $\varepsilon = \min\{|\xi[F^2(\mathbf{x}), F(x)] - \xi[F(x), \mathbf{x}]|/\mathbf{x} \in \mathscr{E}^n, \ \mathbf{x} \ne F^2(\mathbf{x})\}$

Moreover, if $\mathbf{W}$ is an integer matrix or if $\mathbf{W}$ has its elements in $\{-1, 0, 1\}$, the bounds are as before.

This corollary allows us to use the energy for studying the associative memory capabilities of threshold networks [27, 80].

Definition   Let $F\colon \mathscr{E}^n \to \mathscr{E}^n$ be a threshold network, with connection matrix $\mathbf{W}$ and threshold vector $\boldsymbol{\theta}$. Let $\mathbf{X}$ be a matrix with column vectors $\mathbf{X}_k$ in $\mathscr{E}^n$ to be memorised. We will say that F is an *associative memory* for $\mathbf{X}$ iff:

(i) All $\mathbf{X}_k$ are fixed points of the sequential iterations on F.
(ii) All $\mathbf{X}_k$ are attractive in their neighbourhoods:

$$\forall k, \ \exists V_k \subset \mathscr{E}^n \colon \mathbf{X}_k \in V_k$$

$$\forall \mathbf{x}(0) \in V_k, \ \exists T \ge 0 \colon \forall T' \ge T, \ F_\pi^{T'}[\mathbf{x}(0)] = \mathbf{X}_k.$$

According to this definition, the memorised patterns, the $\mathbf{X}_k$, are stable in the recognition process (the iteration), and any pattern sufficiently close to an $\mathbf{X}_k$ will be recognised as this $\mathbf{X}_k$ (see Fig. 7.7). Note that what is intended here is quite similar to what we discussed in Section 7.4.2: we want to retrieve memorised patterns from noisy inputs. But there are two main differences:

(i) The concept of noise is replaced here by the notion of *Hamming distance*: a memorised pattern $\mathbf{X}_k$ must be retrieved from patterns $\mathbf{x}$ 'in a neighbourhood V_k of $\mathbf{X}_k$.
(ii) The retrieval process is *reiterated* until convergence is obtained (fixed point).

Within this formulation it is possible to give conditions that ensure that F is an associative memory for $\mathbf{X}$:

Theorem 5   If $\mathbf{W}$ is a symmetric matrix, with nonnegative diagonal elements, if $\mathbf{W}$, $\boldsymbol{\theta}$ and $\mathbf{X}$ satisfy the following conditions:

$\forall k(1, \ldots, m), \ \forall i, j$:

$$\mathrm{grad}_i E(\mathbf{X}_k) \cdot (X_{ik} - \overline{X_{ik}}) \le 0$$

$$[\mathrm{grad}_j E(\mathbf{X}_k) + (X_{ik} - \overline{X_{ik}})] \cdot [X_{jk} - \overline{X_{jk}}] \le 0$$

then F is an associative memory for $\mathbf{X}$ (where $\bar{0} = 1$ and $\bar{1} = 0$).

The first condition in this theorem can be written as a condition on thresholds when the W_{ij} are given:

$$\forall i(1, \ldots, n): \max\left\{\sum_{j \neq i} W_{ij}X_{jk} / k(1 \ldots m): X_{ik} = 0\right\} < \theta_i \leq$$

$$\min\left\{\sum_{j \neq i} W_{ij}X_{jk} / k(1 \ldots m): X_{ik} = 1\right\}. \tag{7.28}$$

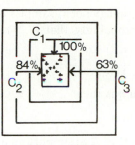

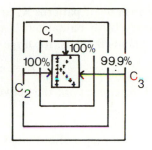

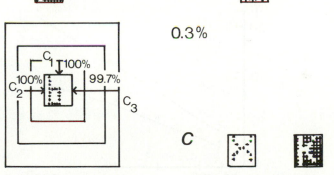

Figure 7.7 Threshold network to retrieve the 3 letters: 'X', 'b', 'K'. The learning rule is Hebb's law with 0-diagonal. The memorised patterns (a) are clearly not independent, which leads to fault tolerance properties being rather weak. The figure (b) shows the attractivity of each memorised item in its three first neighbourhoods C_i (*i* bits different). Spurious attractors appear to be very close to the memorised items, since, for example, 37% of the items in neighbourhood C_3 of 'X' are attracted to a spurious attractor.

Note also that such networks cannot achieve pattern invariance as is shown in (c); an 'X' translated leads to a spurious attractor. Going beyond this limitation requires networks with more complex structure (Section 7.5).

The theorem gives a *necessary and sufficient condition* which ensures that the patterns are fixed points attractive in their first neighbourhood. This condition holds whatever the learning rule used to compute the W_{ij}. This is made possible by appropriately choosing the thresholds θ_i, (see [27] for more details on this), which is different from the approach taken in linear learning, where thresholds are learnt just in the same way as weights.

Condition (7.28) clearly holds for small values of m (e.g. $m = 1$), but, when m increases, it may become infeasible. This allows us to derive a theoretical bound on the storage capacity of the network:

Theorem 6 If patterns $\mathbf{X}_k$ are random, independent samples of a Bernoulli variable with parameter $\frac{1}{2}$, then the probability that (7.28) holds for all thresholds θ_i tends, for large n, to:

$$P = \operatorname{erf}^{nm}[(n - \tfrac{1}{2}(m - 1)]^{1/2} \tag{7.29}$$

and the capacity of the network $c = m/n$ is then:

$$c \sim \tfrac{1}{8} \ln n \tag{7.30}$$

See [27] for a proof of that. Simulations (Fig. 7.8) support relatively well the theoretical expression given in (7.29), especially for large values of n. However, when the patterns are not random (which is generally the case in practice!), the capacity is usually reduced.

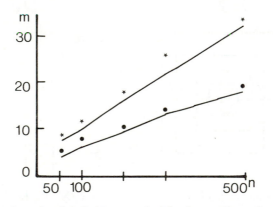

Figure 7.8 Capacity of a threshold network. The figure shows the variation of m, number of random memorised items, as a function of n, size of the network. The solid lines correspond to the theoretical value of m, computed by equation (7.29) with $P = 0.5$, with 0-threshold (bottom) and (top) thresholds computed by:

$$\theta_i = (\tfrac{1}{2})\Sigma_k \Sigma_{j \neq i} [2X_{ik} - 1][2X_{jk} - 1]$$

which is an optimal choice [80]. The simulation results are shown, with the same choice for thresholds, by dots and stars respectively. They are obtained by testing nm inequalities (7.28) per network and adjusting m to get frequency 0.5 of having all inequalities satisfied.

Moreover patterns do *not* need to be independent or orthogonal as was requested in the original work of Hopfield. But then, in the case where **W** is computed by Hebb's rule (7.26), the optimality of **W**, implied by the theory developed in Section 7.3.2, does not hold any more and performances are usually very degraded.

We have represented, in Figs. 7.7 and 7.9, two examples of threshold networks, with matrix **W** computed using Hebb's rule (7.26) and a zero-diagonal constraint, which are used to retrieve letters and words. Note that Hebb's rule is much easier to compute than the usual optimal associative memory matrix given by (7.16). However, performances, in terms of capacity and noise tolerance, are far from being as good when the patterns are not independent and orthogonal.

Another nice feature of the Hebb model compared to the linear model is

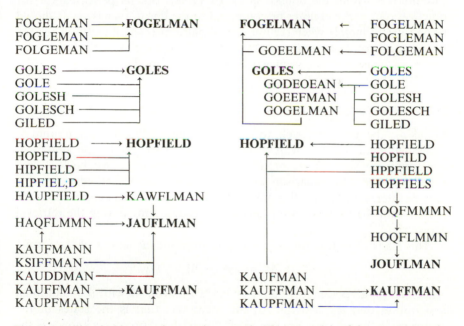

Figure 7.9 Threshold network to retrieve words. The network is of size $n = 104$, each word is coded by concatenating the binary values of the ASCII codes of its letters. We memorised the 4 words: FOGELMAN, GOLES, HOPFIELD, KAUFFMAN, by using Hebb's law with 0 diagonal. But a network of this size is capable of memorising about 8 words. Words in **bold** are fixed points.

Obviously, these words are not (linearly) independent. Nethertheless, our system allows us to retrieve words several bits away from the original word (one letter different is equivalent to at most 8 bits different). This shows that the second condition in our theorem (sufficient for attractivity in the first neighbourhood) is not very restrictive: we did not impose it and the system still works well.

Performances are much improved by working in $\{-1, +1\}$ (left in the figure) instead of $\{0, +1\}$ (right).

that the retrieval process leads to the *exact* stored pattern: stored patterns X_k are made fixed points of the dynamics and thus there is an exact match between the retrieved pattern and the pattern closest to the noisy item (when the distance is not too large). In linear associators, the retrieved pattern is the projection of the noisy pattern on the linear space spanned by the memorised patterns and thus may eventually be non-*identical* to the closest memorised pattern (even with thresholding): linear associators achieve the best possible matching. However, for patterns too far away from a stored item, linear associators still lead to relatively good matching, where threshold networks with Hebb's rule may lead to *spurious* attractors completely different from the memorised items (see Figs. 7.7 and 7.9).

This model can be improved by changing the state space to $\mathscr{E} = \{-1, 1\}$ (which doubles the capacity [80]: see Fig. 7.9), or optimally choosing the thresholds θ within the bounds given by (7.28). But its performances for practical purposes can never be compared to the linear associator. It should be considered mainly when the number of patterns is small and/or when the patterns are preprocessed so as to be made orthogonal: Kohonen also discusses the issue in the framework of the linear model.

7.4.4   *Brain state in the box*

Let us use the term 'feedback' to denote the output Wx to a given input x, computed by a network with connection matrix W. Biological studies in perception – vision, for example – have shown that this feedback is very rapidly evoked, in comparison with the effects of the visualised pattern. It may be interesting to take this observation into account and assume that the pattern and the feedback are present at the same time during the retrieval process.

This leads us to use, for auto-association, a weight matrix of the form:

$$M = \alpha W + \beta I$$

Anderson [5] gives a model based on such a matrix M with the additional assumption that the activities of the 'neurons', that is the states of the automata in the network, are bounded. The retrieval process is known as the 'Brain State in the Box' model (BSB). If the state space is $\mathscr{E} = B = [s, S]$ (B is the 'box'), then for a given input x, the process iteratively computes a sequence $y(i)$ given by:

$$y(0) = x$$

$$y_j(i + 1) = \begin{cases} s & \text{if } M_j y(i) < s \\ M_j y(i) & \text{if } M_j y(i) \in B \\ S & \text{if } M_j y(i) > S \end{cases}$$

where M_j is line j in matrix M.

This iteration procedure is either repeated until a fixed point $\mathbf{y}^*$ is reached or a maximum number of iterations is done: this will be the output of the network. In general the transient time is relatively large and the states y^*_i of the automata are almost all 'saturated', that is either s or S.

7.4.5   *Generalisation*

Let $\mathbf{W} = \mathbf{XX}^+ + \mathbf{Z}(\mathbf{I} - \mathbf{XX}^+)$ be a solution of equation (7.14): $\mathbf{WX} = \mathbf{X}$

The BSB algorithm uses the operator $\mathbf{M} = \alpha\mathbf{W} + \beta\mathbf{I}$ on an original pattern $\mathbf{x}$ inside the box $\mathbf{B}$. This linear operator $\mathbf{M}$ can be written as follows:

$$\begin{aligned}\mathbf{M} &= \alpha\mathbf{W} + \beta\mathbf{I}\\ &= (\alpha\mathbf{XX}^+ + \alpha\mathbf{Z}(\mathbf{I} - \mathbf{XX}^+) + \beta\mathbf{I})(\mathbf{XX}^+ + \mathbf{I} - \mathbf{XX}^+)\\ &= (\alpha + \beta)[\mathbf{XX}^+ + ((\alpha\mathbf{Z} + \beta\mathbf{I})/(\alpha + \beta))(\mathbf{I} - \mathbf{XX}^+)]\\ &= (\alpha + \beta)\mathbf{M}'\end{aligned}$$

The bracketed expression $\mathbf{M}'$ is just a solution of (7.14) with no particular properties. With $\alpha + \beta > 1$, this function $\mathbf{M}$ allows for various iterations, the box limiting the growth of the vectors so that the process will eventually converge (see Fig. 7.10).

This approach can be interpreted in the following way: $\mathbf{M}'$ can be seen as the weight matrix of the network, computed during the learning phase. Let $A_i = \Sigma_j W_{ij} X_j$ be the total input to cell i through the linear transform $\mathbf{W}$. The BSB retrieval process can then be seen as a limiting case of the following retrieval process: $y = f(A_i)$, f being the sigmoid function:

$$f(x) = a(1 - e^{-kx})/(1 + e^{-kx}). \tag{7.31}$$

The size of the box and the slope of the function are respectively defined through the a and k coefficients.

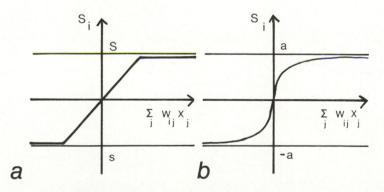

Figure 7.10 Smooth network: a smooth function (right) is given by an expression such as (7.31). The BSB model is represented (left) by the variations of the output S_i in terms of its total input A_i. The variations are limited by the bounds of the box: s, S.

f can be iterated in the same way as with the BSB model, but, usually the fixed point is reached only after a large number of iterations. In our simulations, we stopped the computations after a given number of iterations (usually 10). The parameters *a* and *k* must be optimised to allow for good retrieval performances. A comparison of the two methods performances will be given in Fig. 7.13.

7.4.6   *Simulations*

Simulations were performed on the 26 letters of the alphabet (Fig. 7.11), encoded with -1, 1 values on an $n = 8 \times 8$ grid (corresponding to 64 automata). Three memory sizes have been tested: $m = 10$, 18 and 26 letters. In the following, we present the conditions of the learning and retrieval phases of the experiences.

Learning
The following two algorithms have been tested, both of them with the three memory sizes mentioned above (see Fig. 7.12).

(i) The exact solution of minimal norm of $\mathbf{WX} = \mathbf{X}$, that is $\mathbf{W} = \mathbf{XX}^{+}$, computed through Greville's algorithm (curves W10; W18; W26)
(ii) A constrained solution of this equation (0-diagonal) computed through a projected gradient method (curves $\dot{\mathbf{W}}$10,0; W18,0; W26,0).

In all cases, the desired memorisation has been obtained. But, for the first method, just one presentation of the alphabet was sufficient, where, for method (ii), it was necessary to present the alphabet a number of times which increased with the number of memorised patterns: e.g. 4 presentations of the patterns were needed for $m = 18$ and 10 for $m = 26$.

Figure 7.11 Letters of the alphabet, digitised on an 8 × 8 grid

Retrieval

Three retrieval algorithms have been tested with the 18-patterns case and a 0-diagonal matrix (W18,0):

(i)   Threshold network: the $\mathbf{1}[\mathbf{W}x]$ transform (with 0 as threshold) is iterated until a fixed point is reached.
(ii) BSB network: the function $\mathbf{M} = \alpha\mathbf{W} + \beta\mathbf{I}$ with $\alpha = 0.2$ and $\beta = 0.9$ (as in Anderson [5]) has been iterated inside a $[-1.3, 1.3]^{64}$ box, the number of iterations being bounded to 60. At the end of the process, i.e. when a fixed point has been reached or 60 iterations have been done, the output y is 0-thresholded ($\mathbf{1}[y]$).
(iii) 'Smooth' network: the function $f(x) = a(1 - e^{-kx})/(1 + e^{-kx})$ with $a = 1.7$ and $k = 4$ has been iterated with exactly 10 iterations, and the result ultimately thresholded with 0 threshold.

Figures 7.12 and 7.13 show curves representing the retrieval ratio as a function of the exact number of inverted pixels on memorised patterns. Each point on the curve is an average computed from 100 random trials (of inverted pixels) on each pattern $\mathbf{X}_k$.

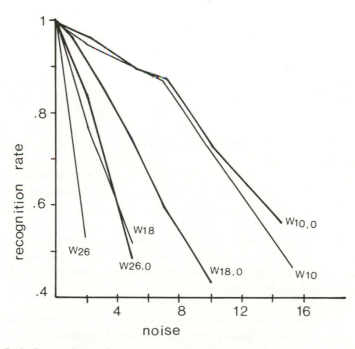

Figure 7.12 Comparison of two learning algorithms

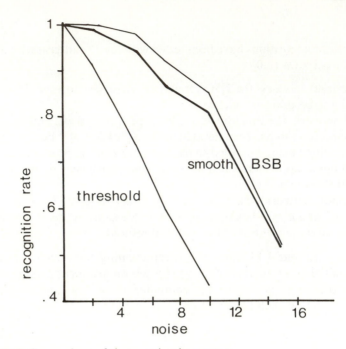

Figure 7.13 Comparison of three retrieval processes

Discussion

The second learning method appears to be better for retrieval performances, whatever the retrieval process, to a degree which increases with the memory size. When the number of memorised patterns is high, XX^+ tends to be the identity so that the input noise is less and less corrected. The minimum norm solution of (7.14) is no longer the best, especially when the input noise is not Gaussian uncorrelated.

During the retrieval, the simple threshold method appeared to allow for only a few iterations and its performances were clearly dominated by the two others. But, as the noise – i.e. number of inverted pixels – increases, so does the number of iteration steps, leading to an increase in the computational burden. This is particularly true for the **BSB** model which needs more iterations than the smooth case; however no computation of an exponential – time-consuming! – is required. A full theoretical analysis of these last two methods is not available at present.

Remark   Letters are always linearly independent. But the recognition rate usually differs between letters and the standard deviation of this rate increases with the number of inverted pixels (see Fig. 7.14).

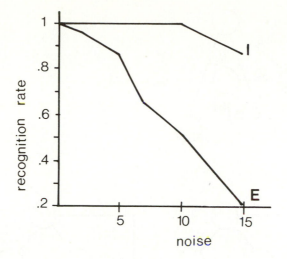

Figure 7.14 Comparison of retrieval performance for letters: 'I ' and 'E'. The performance is clearly better for the 'I'.

7.5 Multilayered networks

7.5.1 *Introduction*

As pointed out in the previous sections, most of the connectionist models are based on the same kind of basic operator: namely a weighted sum of the inputs followed by a nonlinear transformation.

In Section 7.2.3 we introduced the concept of order of a predicate, and we noted that a perceptron could not learn functions of order larger than one, unless it had been specifically designed to do so (Fig. 7.4). We noted that even a boolean function as simple as the exclusive-OR is of order two. This limitation is due to the nature of the threshold function which basically dichotomises its input space with a hyperplane. And it is known that the proportion of linearly separable functions (over all possible boolean functions) goes to zero as the number of input variables increases [15].

Thus, a classification problem of practical complexity has virtually no chance to be directly learnable by a linear classifier. Note that the limitations of linear threshold functions also apply to the associative nets (linear models) described in the previous sections.

Various solutions have been proposed to overcome these limitations:

(i) The first possibility is to parameterise a high-order polynomial hypersurface and to fit its parameters by learning [60]. The limits of this approach are well known (even a simple n-dimensional surface of order 2 has n^2 parameters).

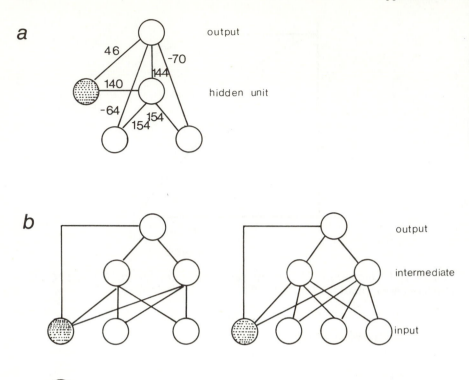

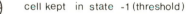

Figure 7.15 Computing *xor* and 3-parity predicate with hidden units. The figure shows various networks with hidden units that compute a *xor*. (a) is a network which was found as the result of the Boltzmann Machine learning algorithm (from [75]). (b) shows networks with two hidden units that compute an *xor* (left) and the 3-parity predicate (right). These networks have been computed using the Gradient Back Propagation algorithm.

Note that the solution found by the Boltzmann Machine algorithm only uses one hidden unit, instead of two for the GBP solution, but connections are different in the two cases. At present, there is no complexity theory about the minimum number of hidden units required.

(ii) Another solution is to precede the learning phase by an adequate 'decoding stage', so as to extract the relevant features before performing the classification. Such an approach has been taken by various authors and proved to perform quite well even on problems of high order [72]. We have seen, for example, that hand-wired feature extractors could allow us to compute the exclusive-OR (Fig. 7.4). The problem with this technique is that the decoding step has to be crafted by hand. It would be most desirable to be able to produce this 'decoder' automatically.

(iii) A solution to this coding problem has been recently proposed by various authors [1, 37, 50, 51, 70], making use of networks with so-called 'hidden units'. Their idea is to take advantage of these units so as to extract *automatically* from the examples the relevant features.

This approach was first developed by Hinton and colleagues [1, 37]: the Boltzmann Machine, a network of probabilistic automata, composed of two sorts of cells (visible and hidden), was the first to demonstrate that it was possible to automatically learn to perform order-2 tasks, such as the exclusive-OR (see Fig. 7.15).

In multi-layered networks, introduced later on [50, 51, 63, 70], the network contains, as in the linear models discussed in the previous sections, an input and an output layer, and moreover intermediate layer(s) which will serve as feature extractors. The cells in these intermediate layers are analogous to the hidden units in the Boltzmann Machine: they have *no* direct interaction with the environment; their state is fully determined by the dynamics on the network itself.

Learning the weights of these internal or 'hidden' units is not straight-forward since no external signal specifies their ideal outputs. So there is no direct way to know whether the output of an internal unit is correct for a particular input pattern or not. This problem has been called the *credit assignment problem* [56].

Let us illustrate the importance of hidden units on a simple example. Consider an automata network as in Section 7.3 (Fig. 7.5) with which we would like to compute an exclusive-OR. This is equivalent to making the following configurations stable points:

$$000xx\ldots x$$
$$011xx\ldots x$$
$$101xx\ldots x$$
$$110xx\ldots x$$

where $xx\ldots x$ is an arbitrary sequence of bits (the same for the four vectors). There is no set of weights (whatever the method used to compute them) that can make these vectors stable without making stable all the other vectors. Each of the first three bits of a line is an XOR of the remaining two. Computing the exclusive-OR or any other nonlinearly separable function with threshold operators may be done by using additional cells. Those cells can compute intermediate predicates such that any output value of a cell can be expressed as a (thresholded) linear combination of the others. For example, the problem just described could be solved by adding a spare unit whose state has been clamped by hand:

$$0000xx\ldots x$$
$$0111xx\ldots x$$
$$1011xx\ldots x$$
$$1101xx\ldots x$$

Now any of the four first bits in the patterns can be computed from the other three through a linear threshold function: in Fig. 7.4, for example, we have represented a network with 1 hidden unit which realises the XOR.

In an attempt to design a general learning rule, the state of the spare units should not be defined by the outside world but should eventually be generated by the network itself, since these bits are not really part of the patterns.

These additional operators are called 'hidden units' [70] because they have no direct interaction with the outside world but are only supposed to cope with the underlying regularities that cannot be taken into account by simple linear pairwise interactions. The concept of hidden (or internal) unit can be used in the case of feed-forward nets (with no loop in the interaction graph) which is a special case of the fully connected net seen in Section 7.4.

We will extensively study multi-layered networks in the following sections. In this case, the hidden units compute intermediate representations that decompose the input-output function into elementary successive steps (see Fig. 7.16). However, learning these intermediate predicates without requiring the environment (or the designer) to explicitly specify them is not a simple task, because what is addressed here is the problem of *generating internal representations*. It can be compared, in a different context, to the problem of decomposing a complex task (or a function) into simpler subtasks (or elementary operators).

The credit assignment problem was an unsolved problem until recently. Some authors [56] even suggested in the late 1960s that such generalisations of the neural network paradigm might not exist. This provides a possible explanation for the decrease of interest in the field during the 1970s.

Its recent renewal originated from the learning procedures recently proposed for the *Boltzmann Machine* [1] and for multi-layered feed-forward networks. This last algorithm, called *Gradient Back Propagation* (GBP), was developed by several authors independently [50, 51, 70]. It is based on the

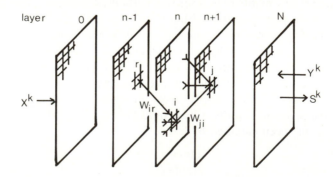

Figure 7.16 Multi-layered network with $N - 1$ intermediate layers. Only the connections between successive layers are allowed. Some weights are shown.

propagation of error signals and solves the credit assignment problem. In the following, we will describe this last procedure, using intuitive arguments. We will then give a general theoretical method for deriving such algorithms and finally give examples of applications.

7.5.2 *GBP: a learning algorithm for multi-layered networks*

This section will be concerned with networks composed of several layers of neuron-like processing elements.

The first one (number 0) is the input layer, its state (the input vector) will be denoted $\mathbf{X}^h$. The last layer (number N) is the output layer, its state, the output vector will be denoted $\mathbf{S}^h$.

We will assume for the sake of simplicity that the connections can only go from lower layers to higher layers, and that there is no connection between cells in the same layer. Connections can skip layers.

The elements compute a weighted sum of their inputs, followed by a nonlinear transformation f. Let $\mathcal{X}_i$ be the output of cell i:

$$\mathcal{X}_i = f[\Sigma_j W_{ij} \cdot \mathcal{X}_j] \qquad (7.32)$$

where W_{ij} is the weight between cell j and cell i. As usual we will define A_i, the *total input* to cell i, as:

$$A_i = \Sigma_j W_{ij} \cdot \mathcal{X}_j \qquad (7.33)$$

such that we can now write:

$$\mathcal{X}_i = f[A_i] \qquad . (7.34)$$

Here, f is a *differentiable function*. In the following, it will be assumed that f is odd, with a strictly positive first derivative (e.g. a sigmoid function: see Fig. 7.8). At time h, one element $\mathbf{X}^h$ of the family of input patterns $\mathbf{X}_k$, is presented on the first layer, the stable state of the network is computed by propagating the information through the layers using equation (7.34). Simultaneously, a desired output vector $\mathbf{Y}^h$ is given to the cells of the last layer (see Fig. 7.14). In the previous sections, we have presented some learning procedures for finding the weights of a formal neuron from examples of desired input-output pairs. All these methods are based, in an implicit or explicit way, on the iterative minimisation of a particular criterion attached to the automaton. In the simplest case, this criterion is the squared difference between the computed output and the desired output of the unit. In a multi-layer network, only the units belonging tof the last layer are given a desired output.

For those cells, a criterion C_k measuring the discrepancy between the desired output $\mathbf{Y}_k$ and the actual output $\mathbf{S}_k$ of each output unit for the kth input pattern $\mathbf{X}_k$, is defined:

$$C_k = \sum_s [S_{sk} - Y_{sk}]^2 \tag{7.35}$$

where the sum is taken over the units, indexed by s, of the output layer only.

S_{sk} thus corresponds to the state $\mathscr{X}_s$ of cell s in the output layer. We now define a global criterion $\mathscr{C}$ as the sum of the C_k, the sum being taken over the m possible input-output pairs:

$$\mathscr{C}(\mathbf{W}) = \frac{1}{m} \sum_k C_k \tag{7.36}$$

Instead of minimising $\mathscr{C}$ directly, we will minimise an estimate of $\mathscr{C}$ using the stochastic gradient method similar to the Widrow–Hoff rule described in Section 7.3.4. From now on, for notational convenience, we will skip the temporal index h: C will denote the cost associated with *one* pattern.

In order to modify the weights using gradient descent, we have to calculate the effect of their variations on the value of the criterion, that is the partial derivatives of the criterion with respect to these weights: $\partial C / \partial W_{ij}$. Rather than computing all those derivatives directly, we will first compute the gradient of the criterion relative to the total input of each cell: $\partial C / \partial A_i$. Once these derivatives are known, the derivatives relative to the weights are easily computed with the following formula (given by the definition of A_i):

$$\partial C / \partial W_{ij} = \partial C / \partial A_i \cdot \mathscr{X}_j \tag{7.37}$$

that enables us to apply the gradient descent procedure.

The $\partial C / \partial A_s$ associated to the output cells immediately follow from equations (7.32) and (7.35):

$$\partial C / \partial A_s = 2[S_s - Y_s] \cdot \partial S_s / \partial A_s$$

which can be rewritten using (7.34):

$$\partial C / \partial A_s = 2[S_s - Y_s] \cdot f'(A_s) \tag{7.38}$$

The computation of the $\partial C / \partial A_i$ associated to the hidden cell i can be easily performed, provided that all the $\partial C / \partial A_j$ of the cells j to which cell i sends its output are known. We simply state (using the chain rule):

$$\partial C / \partial A_i = \sum_j \partial C / \partial A_j \cdot \partial A_j / \partial A_i$$

We now need to compute $\partial A_j / \partial A_i$:

$$\partial A_j / \partial A_i = \partial A_j / \partial \mathscr{X}_i \cdot \partial \mathscr{X}_i / \partial A_i$$

The first factor on the right-hand side is equal to W_{ji}, while the second is $f'(A_i)$. Combining the two previous equations gives:

$$\partial C / \partial A_i = f'(A_i) \cdot \sum_j W_{ji} \cdot \partial C / \partial A_j$$

For clarity, we will denote, in the following, the gradient of C with respect to A_i by:

$$\mathcal{Y}_i = \partial C / \partial A_i \qquad (7.39)$$

The previous equation and equation (7.38) then respectively become:

$$\mathcal{Y}_i = f'(A_i) \cdot \sum_j W_{ji} \cdot \mathcal{Y}_j \qquad (7.40)$$

$$\mathcal{Y}_s = 2[S_s - Y_s] \cdot f'(A_s) \qquad (7.41)$$

Equation (7.40) provides a mean for computing the gradient of the cost C with respect to the total input of a cell by performing the weighted sum of the $\mathcal{Y}_s$ corresponding to the cells to which it sends an output. The weights involved in this operation are the same as in the forward pass, but used backwards. Thus in a multi-layered feed-forward net, the $\mathcal{Y}_s$ are back-propagated from the last layer to the first one. Note that despite the fact that this section was primarily concerned with feed-forward networks, the arguments developed here can be easily transposed to any network (with loops).

In short, one iteration of the back propagation learning algorithm consists in (see Fig. 7.17):

(i) Present pattern $\mathbf{X}^h$, and compute the state $\mathcal{X}$ of the network by forward propagation (equation (7.34)):

For an input cell i $\qquad \mathcal{X}_i = X^h_i$

From layer 1 to layer N

$$\mathcal{X}_i = f(A_i) \quad A_i = \Sigma_j W_{ij} \mathcal{X}_j$$

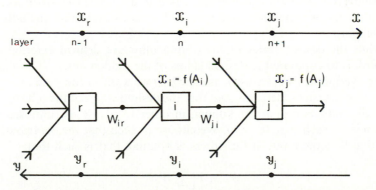

Figure 7.17 Gradient Back Propagation algorithm. The figure shows the forward pass (top) to compute the states $\mathcal{X}$, from layer 1 to layer N, and the backward pass (bottom) to compute the gradients $\mathcal{Y}$, from layer $N - 1$ to layer 1.

(ii) Present the desired output $\mathbf{Y}^h$, compute the gradients $\mathcal{Y}_s$ on the last layer using equation (7.4.1), and compute the $\mathcal{Y}_i$ corresponding to the hidden units by back-propagation (7.40).

For an output cell s $\qquad\qquad \mathcal{Y}_s = 2[S_s - Y^h_s] \cdot f'(A_s)$

From layer $N-1$ to layer 1 $\quad \mathcal{Y}_i = f'(A_i) \cdot \Sigma_j W_{ji} \cdot \mathcal{Y}_j$

(iii) Modify the weights by inserting (7.37) in a stochastic gradient method:

$$W^h_{ij} = W^{h-1}_{ij} - \lambda^h \mathcal{Y}_i \mathcal{X}_j \qquad\qquad (7.42)$$

7.5.3 *Simplified GBP algorithm: the Hierarchical Learning Machine*

Several back-propagation procedures for learning the weights of a multi-layer network had been proposed before a theoretical formulation appeared.

The Hierarchical Learning Machine algorithm (HLM) [50, 51] is one of them. The HLM learning procedure is basically an approximation of the GBP in the case of binary units (threshold functions: $f = \mathbf{1}$), that uses as criterion:

$$C_k = \sum_s [Y_{sk} - A_s]^2$$

where A_s is the total input to cell s in the last layer, when the input is $\mathbf{X}_k$.

Then, it is easy to see that the back-propagated variables are not error signals but desired states. To keep the formula comparable, we will use, in this section, the same symbol $\mathcal{Y}$ for the desired states as for the gradients in the previous section.

Instead of a full mathematical development, which can be found in [50, 51], we will just describe here the basic idea of the HLM learning algorithm. The simplest learning strategy is to use a Widrow–Hoff-like rule, provided that we can associate a desired state to each unit, including the hidden units.

While the desired states of the output units are defined externally, the problem is to compute the desired states of the hidden units. The following informal argument will be used: assume that we want to compute the desired state $\mathcal{Y}_i$ of the ith hidden unit, and that the desired states of the cells to which this cell sends its output are known; in other words, $\mathcal{Y}_j$ is supposed to be known if j is such that $W_{ji} \neq 0$. We choose $\mathcal{Y}_i$ such that the contribution of the ith cell to the input of the others is optimal, that is such that:

$$\mathcal{Y}_i = \mathbf{1}\left(\sum_j W_{ji} \mathcal{Y}_j \right) \qquad\qquad (7.43)$$

which states that the desired output of cell i is the binary value ($+1$ or -1) nearest to the average of the desired states of the following cells. The sum is weighted with the W_{ji} because we want to get a cell i, followed by a cell j, to fall into a desired state that 'satisfies' cell-j state in proportion to weight W_{ji} (the larger W_{ji}, the more likely $\mathscr{Y}_i = \mathscr{Y}_j$). Compare equation (7.43) with equations (7.11) – perceptron criterion – and (7.40) – back propagation of gradients.

A learning iteration is composed of the three following phases:

(i) Present pattern $\mathbf{X}^h$, and compute the state $\mathscr{X}$ of the network by forward propagation:

for an input cell i: $\quad \mathscr{X}_i = X^h_i$

otherwise: $\qquad\qquad \mathscr{X}_i = \mathbf{1}(A_i) \quad A_i = \Sigma_j W_{ij} \mathscr{X}_j$

(ii) Present the desired output $\mathbf{Y}^h$ on the last layer, and compute the $\mathscr{Y}_i$ corresponding to the hidden units by back-propagation.

otherwise:
for an output cell s: $\quad \mathscr{Y}_s = Y^h_s$

$$\mathscr{Y}_i = \mathbf{1}(\Sigma_j W_{ji} \mathscr{Y}_j)$$

(iii) Modify the weights using the standard Widrow-Hoff rule:

$$W^h_{ij} = W^{h-1}_{ij} - \lambda^h(A_i - \mathscr{Y}_i)\mathscr{X}_j$$

This procedure can be derived in a more formal way from the criterion defined above. This algorithm is less stable than the 'continuous GBP' but it is simpler and exhibits some interesting properties (see [50, 51]).

7.5.4 *The GBP algorithm as a constrained minimisation process*

A general framework for deriving GBP-type procedures is presented in this section. Extensive use of the *Lagrangian* formalism is made in a context similar to optimal control theory [6]. For the sake of clarity, this method will be described in the special case of multi-layered feed-forward networks where connections cannot skip layers.

Consider a network with $N + 1$ layers, the state vector of the kth layer is denoted $\mathscr{X}(k)$, the corresponding weight matrix, connecting the $(k - 1)$th layer to the kth layer is $\mathbf{W}(k)$. The global state is thus defined by:

$$\mathscr{X}(k + 1) = \mathbf{F}[\mathbf{W}(k + 1)\mathscr{X}(k)]$$

or with the usual notation:

$$\mathscr{X}(k + 1) = \mathbf{F}[\mathbf{A}(k + 1)] \quad \text{with} \quad \mathbf{A}(k + 1) = \mathbf{W}(k + 1)\mathscr{X}(k) \quad (7.44)$$

with $\mathscr{X}(0) = \mathbf{X}^h$, and $\mathbf{F}$ the function with all its components equal to f.

Let $\mathbf{Y}$ be the desired state of the last layer. We define the *Lagrange* function L:

$$L = [\mathbf{Y} - \mathscr{X}(N)]'[\mathbf{Y} - \mathscr{X}(N)]$$
$$+ \Sigma_{k=0\ldots N-1}\mathbf{Z}'(k + 1)\cdot[\mathscr{X}(k + 1) - \mathbf{F}[\mathbf{W}(k + 1)\mathscr{X}(k)]] \quad (7.45)$$

The first term in this equation is the usual quadratic cost function to be minimised. (We can consider that L depends only on one pattern pair $\mathbf{X}$, $\mathbf{Y}$ without loss of generality, since L is additive.)

The second term represents the constraints of the problem, $\mathbf{Z}(k)$ being the Lagrange multipliers. The constraints terms are equal to 0 if equation (7.44) is satisfied. The optimality conditions that describe the behaviour of the network are:

$$\partial L/\partial Z_i(k) = 0 \quad \text{for all } i \text{ and } k$$
$$\partial L/\partial \mathscr{X}_i(k) = 0 \quad \text{for all } i \text{ and } k$$
$$\partial L/\partial W_{ij}(k) = 0 \quad \text{for all } i, j \text{ and } k \quad (7.46)$$

The first condition simply gives $\mathscr{X}(k + 1) = \mathbf{F}[\mathbf{A}(k + 1)]$ for $k = 0, 1, \ldots,$ $N - 1$, which describes the forward pass.

The second condition gives two different equations depending whether $k = N$ or not. If $k = N$ we obtain:

$$\mathbf{Z}(N) = 2[\mathbf{Y} - \mathscr{X}(N)] \quad (7.47)$$

If $k \neq N$ we have:

$$\mathbf{Z}(k) = \mathbf{W}'(k + 1)\cdot\mathbf{D}_F[\mathbf{A}(k + 1)]\cdot\mathbf{Z}(k + 1) \quad (7.48)$$

where $\mathbf{D}_F[\mathbf{A}(k + 1)]$ is the Jacobian matrix of $\mathbf{F}$ at point $\mathbf{A}(k + 1)$. In our case, this matrix is diagonal, and its ith term is equal to $f'[A_i(k + 1)]$. For clarity, the previous equation can be rewritten in scalar form:

$$Z_i(k) = \Sigma_j W_{ji}(k + 1)\cdot f'[A_j(k + 1)]\cdot Z_j(k + 1)$$

By denoting $\mathscr{Y}_i = f'[A_j(k + 1)]\cdot Z_j(k + 1)$ and substituting in (7.47) and (7.48) we obtain:

$$\mathscr{Y}(N) = 2\mathbf{D}_F[\mathbf{A}(N)][\mathbf{Y} - \mathscr{X}(N)]$$
$$\mathscr{Y}(k) = \mathbf{D}_F[\mathbf{A}(k)]\mathbf{W}'(k + 1)\mathscr{Y}(k + 1) \quad (7.49)$$

which describe the backward pass in the GBP algorithm.

The third condition in (7.46) does not give a practical method for computing the weights but just gives an optimality condition:

$$\mathscr{Y}(k)\mathscr{X}'(k - 1) = 0 \quad \text{for all } k$$

However, the weights that satisfy this condition can be easily computed using a gradient descent algorithm:

$$W_{ij}^{h+1}(k) = W_{ij}^h(k) - \lambda^h\cdot\partial L/\partial W_{ij}^h(k)$$

This gives:

$$W_{ij}^{h+1}(k) = W_{ij}^h(k) + \lambda^h\cdot\mathscr{Y}_i(k)\mathscr{X}_j(k - 1) \quad (7.50)$$

or using matrix notation:

$$\mathbf{W}^{h+1}(k) = \mathbf{W}^h(k) + \lambda^h \cdot \mathscr{Y}(k)\mathscr{X}'(k-1) \tag{7.51}$$

These results can be summarised by the following system:

(i) For $k = 0, 1, \ldots, N-1$

$$\mathscr{X}(k+1) = \mathbf{F}[\mathbf{A}(k+1)] \quad \text{with} \quad \mathbf{A}(k+1) = \mathbf{W}(k+1)\mathscr{X}(k)$$

(ii) For $k = N-1, N-2, \ldots, 1$

$$\mathscr{Y}(k) = \mathbf{D}_F[\mathbf{A}(k)]\mathbf{W}'(k+1)\mathscr{Y}(k+1)$$

(iii) For $k = 0, 1, \ldots, N-1$

$$\mathbf{W}^{h+1}(k+1) = \mathbf{W}^h(k+1) + \lambda^h \cdot \mathscr{Y}(k+1)\mathscr{X}'(k)$$

(iv) And the 'boundary conditions':

$$\mathscr{Y}(N) = 2\mathbf{D}_F[\mathbf{A}(N)][\mathbf{Y}^h - \mathscr{X}(N)]$$

$$\mathscr{X}(0) = \mathbf{X}^h$$

This system is identical to classical GBP, except that the output of a layer depends exclusively on the outputs of the previous one. However, a similar derivation can be made without this assumption.

Many different Lagrange functions can be chosen, each of them leading to a different learning algorithm which depends on the choice of the cost function, the nonlinear transformation $\mathbf{F}$, and the particular architectural assumptions. For example, L can be chosen such that $\mathbf{W}$ does not depend on the layer index k; in that case, the matrix connecting a layer to the next one is the same for every layer. The index k can now be interpreted as a time index, and the Lagrange function can be viewed as describing the time evolution of a cross-connected network. The learning rule involves an initial state $\mathbf{X}$ at time $k = 0$, and a final (target) state $\mathbf{Y}$ at time $k = N$. The rule is identical to the one described in [70] for pseudo-iterative nets and depends on the sum of the gradients over the trajectory.

7.5.5 *Experiments with the GBP algorithm*

The dynamic behaviour of the GBP algorithm can be analysed using simple learning tasks. We propose in this section some typical examples.

As shown below, a multi-layer network can be trained to compute boolean functions such as the exclusive-*or*, the *n*-parity problem, or the product of two 3-bit binary numbers. The function f chosen in our simulations is:

$$f(x) = a[1 - e^{-kx}]/[1 + e^{-kx}] \tag{7.52}$$

This function is odd, strictly increasing, and varies continuously between $-a$ and $+a$ (see Fig. 7.10).

Synthesis of boolean functions

The problem here is to synthesise a network that computes a boolean function defined by its truth table. This is not strictly speaking a learning task, since we do not expect any generalisation: it is just rote learning. However, besides memorising the truth table, the network has also built in its weight structure a *compact* representation of the function. This is similar to the classical problem in electronic circuit design of decomposing a complex function into a small set of basic components.

We have seen before (see Fig. 7.4) that a perceptron cannot compute an exclusive-*or* unless specifically hand-wired. In contrast, it is very easy to build a multi-layer network that learns how to perform this task: the input layer has two units, the output layer one, and the intermediate layer may have only one hidden cell. Of course, various architectures are possible. We present in Fig. 7.14 an example of such a network. Obviously, the solution in weight space is not unique: there exist many decompositions of the exclusive-*or* function into threshold functions, which can be obtained from various initial conditions on W and computing the 'threshold equivalent' of the smooth functions obtained (see Fig. 7.3).

The *n*-parity predicate (see Section 7.3.) can be learned by a similar network, but with a number of hidden units which depends on n (see Fig. 7.15).

It is also possible to propose a network that learns to compute the product of two 3-bit binary numbers. The first layer, with two groups of 3 units encodes the numbers to be multiplied, the second one has 30 hidden cells, and the last one outputs the result, in binary form, on 6 cells. There is a full connectivity between cells in successive layers (see Fig. 7.18).

With only 40 presentations of the 64 possible inputs, the network achieves 100% success.

Associative memory

We will present here some results of simulations intended to compare the performance on an auto-association task of various algorithms discussed so far. The patterns to be memorised are characters X_k coded on 8×8 bit $(-1$ or $+1)$ matrices. We use the first 18 letters of the alphabet, digitised by hand (see Fig. 7.11).

We use the following learning algorithms:

(1) The auto-associative model with a matrix W computed from (7.20) (Gram–Schmidt orthogonalisation) and a constraint of O-diagonal (Section 7.35) on W (Z is thus nonzero).

(2) An auto-associative two-layer network, with matrix W computed by a Widrow–Hoff rule and O-diagonal constraint, on a criterion similar to (7.35):

$$C(W) = \frac{1}{m} \Sigma_k C_k$$

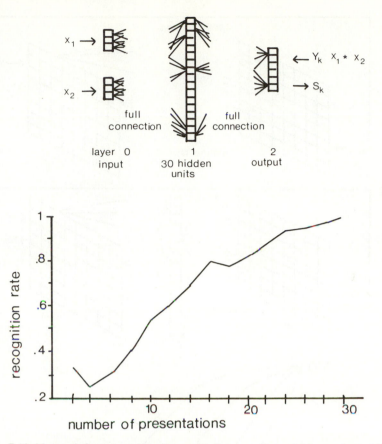

Figure 7.18 Network for computing the multiplication of two 3-bit numbers. (a) the architecture of the network, (b) the recognition rate as a function of the number of presentations of the full set of examples.

with: $$C_k = \Sigma_s [f(\Sigma_j W_{sj} X_{jk}) - X_{sk}]^2$$

where f is a sigmoid function as in (7.52), with $a = 1.7$ and $k = 1.3$ (which implies that: $f(1) \sim 1$).

(3) The same network as before but with a different learning strategy. During the first part of the learning session, for each pattern X_k, various noisy versions $\underline{X}_k$ (the pattern X_k where 5 bits, chosen randomly, have been inverted) are presented to the network. This means that, instead of associating the X_k to itself, we first associate noisy versions of X_k (the $\underline{X}_k$) to X_k; the input pattern $\underline{X}_k$ is thus associated to the desired output X_k. In a second phase, when the weights are stabilised, the 'pure' patterns X_k are stored in the network, ensuring that they will be correctly memorised. Formally, this looks like a hetero-association process, but in fact gives a constrained solution to the auto-association problem.

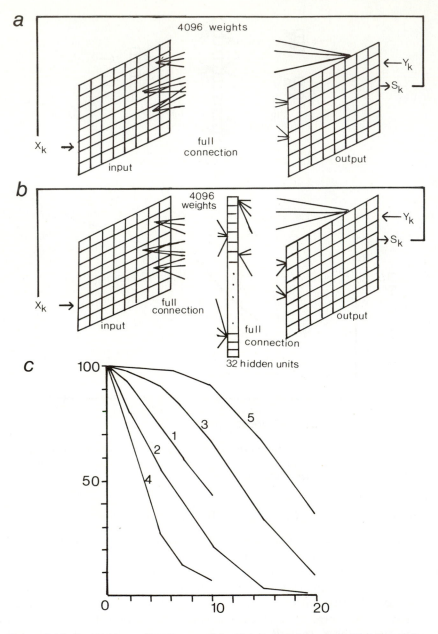

Figure 7.19 Comparison of various models of associative memory. The figure shows the architectures of various networks used for associative memory. (a) Two-layered network, used in cases 1, 2 and 3 (see text). (b) The three-layered network, used in cases 4 and 5. (c) The rate of correct retrieval as a function of the distance (in bits) of the input pattern to the memorised item, showing the five curves associated to the five learning models discussed in the text.

(4) a multi-layer network, with the same number of weights (4096) as the previous two-layered networks: the input and the output layers are the same (64 cells each), the intermediate layer has 32 hidden units. There is full connectivity among successive layers. We use the Back Propagation Algorithm with the same learning strategy as in (2).
(5) The same network as in (4), with the learning strategy of (3), that is 3-layer network and noisy auto-association.

The retrieval process is an iterated thresholding of the output of the network computed by (7.52): that is the output computed through the smooth function is thresholded ($+1$ if positive, -1 otherwise) and the result is fed back into the network at the input layer. This process is repeated for 5 iterations. Figure 7.19 shows, in the five different cases listed above, the curves giving the success rate (percentage of correct retrieval of pattern $\mathbf{X}_k$ from a noisy pattern $\underline{\mathbf{X}}_k$) as a function of the Hamming distance from X_k to $\mathbf{X}_k$. Each point in the curves is an average over 50 samples of patterns X_k for each of the 18 patterns $\mathbf{X}_k$.

We observe that:

Performances are always better when a 'model of noise' has been given to the network (through the examples). The purpose of the retrieval is to cancel the noise present in the input patterns. There is no reason why there should exist a universal solution, independent of the type of noise under consideration. We have shown, in the linear framework (Section 7.4.2), that for each model of noise there exists an appropriate choice for $\mathbf{W}$. The learning strategy used in cases (3) and (5) can be viewed as a method for providing the network with an implicit model of noise, which constrains the choice of $\mathbf{W}$.

However, we have seen that the linear learning model (curve 1) has an inherent model of Gaussian uncorrelated noise (Section 7.4.2). The Widrow–Hoff rule for two-layered networks, which does not find the corresponding minimum norm solution, can nevertheless capture some model of noise, though in a very limited way (curves 2 and 3).

In contrast, the multi-layer model does not have any built-in assumption on noise. This explains the poor performances obtained in case (4) which can be easily fixed, as demonstrated by curve (5), by just providing the network with a model of noise during learning.

The multi-layer architecture clearly achieves better performances, which proves that it is more able to capture the regularities of the environment, storing them in the weights in a more efficient way.

It must be stressed that these kinds of problems are very under-constrained. Providing some information on the noise thus helps the constraining of the set of solutions, hence finding a more adapted solution.

The learning processes described above can be interpreted as moving the weight vector on a high-dimensional error surface. The shape of this error surface determines the complexity of the task to be learned.

When there is no hidden unit, this surface is a paraboloid. Since it is convex, the steepest descent procedure always finds an optimal solution for any initial weight. It can frequently happen that the solution is degenerate, that is an entire subspace in the weight space is solution to the problem. In that case, the particular solution found by the gradient descent, depends on the initial weight configuration.

For multi-layer networks (with hidden units), the situation is very different since the error surface is generally *not convex*. This has several important implications. First, the final solution depends on the initial weight configuration; second, the order of presentation of the patterns, as well as their relative frequencies of occurrence, influence the result by modifying the shape of the error surface locally in time. The third consequence is that the weights can be trapped in a local minimum of the error surface.

tic-tac-toe

The game of tic-tac-toe is a 2-player game, played here on a 3×3 board. Each player plays at his turn by writing a cross (for player 1) or a circle (for player 2). The first one who succeeds in getting 3 figures in line wins. We present a multi-layered network which learns to classify the configurations of the game into win, loss or draw (we always take the viewpoint of the player who plays cross and assume that players play their best strategy).

The network has 3 layers: the input layer codes the board configuration (i.e. a vector with 27 elements: 3 groups of 3×3 cells coding the state of the cell on the board), the output layer has 3 cells, coding the outcome corresponding to the configuration (i.e. a vector with 3 components; win, loss, draw). The intermediate layer, with 50 cells, is fully connected, to both the input and output layers.

We use a learning set containing half of the 5478 legal configurations and test the results of the learning process on the remaining half. The recognition rate (Fig. 7.20) steadily increases with the number of presentations of the examples: after 10 such presentations, the learning rate is about 0.9 and the generalisation rate 0.76 (the learning rate measures the percentage of correct answer for an input in the learning set; and the generalisation rate, the percentage of correct answers for an input in the test set).

Of course, those rates depend on the number of presentations of the examples, on the number of hidden units and on the parameter λ^h used in the learning rule (7.42). Note also that those performances have been obtained with the 'standard' network and learning algorithm. No *a priori* knowledge about the domain has been used to improve performances.

7.5.6 Conclusion

The Gradient Back Propagation algorithm presented in this section provides an answer to the problem of the limitation of all linear associator-type machines. It can compute high-order predicates, through an algorithm that is universal (i.e. does not depend on the domain). The learning session allows the network to elaborate intermediate predicates that decompose the original task into elementary simple steps. The resulting representation of the characteristic features of the examples allows the network to generalise to new input patterns. However, the learning session must take into account samples of the possible inputs, that is the learning set must include a 'model of noise' (Fig. 7.19).

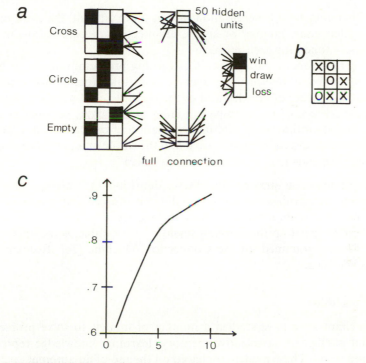

Figure 7.20 Network for the tic-tac-toe game.
The figure shows a network that learns to tell winning configurations of a game from null or losing.
(a) The structure of the network: 3 blocks of 3 × 3 cells receive the input (the configuration of (b) is coded on the input layer). The intermediate layer has 50 hidden cells and the output layer 3 cells which code the output: the configuration shown in (b) is winning, so the output 100 is shown. There is complete connection between the input layer and the intermediate layer, and between the intermediate layer and the output layer.
(b) An admissible configuration.
(c) The learning rate as a function of the number of presentations of the examples.

Let us end this section with some general remarks on the algorithms we have proposed:

Increasing the number of weights usually improves performance, but may drastically decrease the generalisation rate: the network, when possible, just performs rote learning. Hence, a critical balance must be kept, depending on the task:

To achieve rote learning (as in the problem of automatic synthesis of boolean functions), a minimum number of hidden units is necessary (no 'complexity' theory to determine this number is available) to perform the learning. Increasing the number of units usually helps, but varying the architecture (number of layers, connection structure) may have similar effects.

Obtaining good generalisation rates is more difficult: the exact number of hidden units, number of layers, architecture of connections must be precisely tuned for each problem.

There are many parameters in a learning problem: number of hidden units, number of intermediate layers, connection structure between layers, pedagogy (order of presentation of the examples and number of present-ations), choice of the examples, coding of the data, parameters of the learning algorithm (λ_h and initial distribution of weights). Today, all these parameters have to be hand-crafted and the whole outcome of the learning session depends heavily on the options taken.

Finally, it must be stressed that these algorithms are always very time-consuming. A learning session for small-sized problems may take hours of computation. But implementing those techniques on parallel machines may considerably speed up the learning session: for example, some applications have been programmed on the Connection Machine [36]. Recognition is always very fast.

7.6 Conclusion

In this chapter we have studied a number of models to solve problems in artificial intelligence: associative memory, learning, knowledge representa-tion, and so on. These models are based on the use of an automata network. Its connection weights are produced automatically by a learning rule that extracts the relevant information from examples presented to the network. The task is thus decomposed into a learning session, when examples are presented to the network, and a retrieval session, when new patterns are shown to the network which is supposed to process them, using the internal representations elaborated during the learning session.

The learning algorithms presented here can be classified in two different families: the linear-type (such as perceptron and linear associator) and the multi-layer type (such as GBP or Boltzmann Machine). The performances of these families are quite different: the linear family can perform only low-order computations, whereas the multi-layered networks have been shown to perform well on a variety of more complex tasks.

The Gradient Back Propagation, proposed only two years ago, is not yet fully understood. Its properties must be explored in more detail to elucidate in particular:

the exact role of the different parameters,
the minimum requirements in terms of number of cells,
the kind of generalisation that takes place,
the relationship between this type of learning and the classical logic-based learning.

This line of research has already proved on 'toy' examples that it is able to provide interesting learning capabilities. It remains now to test the model on real-world applications, which will help us understand its limits.

A characteristic of this field – as of the general field of automata – is that many authors from various scientific domains have contributed to it, often independently: physicists, neurophysiologists, mathematicians, computer scientists, psychologists. Much work remains to be done and tools from these different areas might prove useful.

7.7 References

[1] D. H. Ackley, G. E. Hinton and T. J. Sejnowski, A learning algorithm for Boltzmann Machines, *Cognitive Science*, 9 (1985), 147–69
[2] S. I. Amari. Learning patterns and patterns sequences by self-organizing net of threshold elements. *IEEE Trans. Comput.*, C-21, 11 (Nov. 1972)
[3] J. A. Anderson, *The Architecture of Cognition*, Harvard University Press, Cambridge, Mass. (1983)
[4] J. A. Anderson, Cognitive and psychological computation with neural models, *IEEE Trans. Syst. Man. & Cybern.*, SMC 13 (1983), 799–815
[5] J. A. Anderson, Cognitive capabilities of a parallel system, in [12], 209–226.
[6] M. A. Arbib and A. R. Hanson (eds), *Vision, Brain and Cooperative Computation*, MIT Press, Cambridge, Mass. (1985)
[7] M. Athans and P. L. Falb, *Optimal Control*, McGraw-Hill, New York (1966)
[8] D. H. Ballard, Parameter nets, *Artificial Intelligence*, 22 (1984), 235–67
[9] D. H. Ballard and C. M. Brown, *Computer Vision*, Prentice-Hall, Englewood Cliffs, NJ (1982)
[10] D. H. Ballard, G. E. Hinton and T. J. Sejnowski, Parallel visual computation, *Nature*, 306, 3 (1983) 21–6
[11] A. Barr, P. R. Cohen and E. A. Feigenbaum, *The Handbook of Artificial Intelligence*, Vol. 3, Pitman, London (1982)
[12] E. Bienenstock, F. Fogelman Soulié and G. Weisbuch (eds), *Disordered Systems and Biological Organization,* Springer Verlag, NATO Asi Series in Systems and Computer Science, n°F20, Berlin, Heidelberg, New York (1986)
[13] S. L. Campbell and C. D. Meyer, *Generalized Inverses of Linear Transformations*, Pitman, London (1979)
[14] J. P. Changeux, *L'Homme Neuronal*, Fayard, (1983)

[15] T. M. Cover, Geometrical and statistical properties of systems of linear inequalities with applications to pattern recognition, *IEEE Trans. Electronic Computers*, EC 14, 3 (1965), 326–34

[16] J. Demongeot, E. Goles and M. Tchuente (eds), *Dynamical Systems and Cellular automata*, Academic Press, New York (1985)

[17] R. O. Duda and P. E. Hart, *Pattern Classification and Scene Analysis*, Wiley, New York (1973)

[18] G. M. Edelman and G. N. Reeke, Selective networks capable of representative transformations, limited generalizations and associative memory, *Proc. Natl. Acad. Sci. USA*, 79 (1982), 2091–5

[19] D. Farmer, T. Toffoli and S. Wolfram (eds), *Cellular Automata*. Physica 10D, North-Holland, Amsterdam (1984)

[20] J. A. Feldman and D. H. Ballard, Connectionist models and their properties, *Cognitive Science*, 6 (1982), 205–54

[21] J. A. Feldman, Dynamic connections in neural networks, *Biol. Cybern.*, 46 (1982), 27–39

[22] J. A. Feldman, Connections, *Byte* (April 1985), 277–83

[23] F. Fogelman Soulié, Contribution à une théorie du calcul sur réseaux, Thesis, Grenoble (1985)

[24] F. Fogelman Soulié: Cerveau et machines: des architectures pour demain? '*Cognitiva 85*', CESTA-AFCET (ed.), actes du forum, (1985) in French.

[25] F. Fogelman Soulié, P. Gallinari and S. Thiria, Linear learning and associative memory. In *Pattern Recognition, Theory and Applications*, J. Devijver (ed.), NATO ASI Series in Computer Science, Springer Verlag, Berlin, Heidelberg, New York to appear.

[26] F. Fogelman Soulié, E. Goles-Chacc: Knowledge representation by automata networks, in *Computers and Computing*, P. Chenin, C. di Crescenzo, F. Robert (eds), Masson-Wiley (1986), 175–80

[27] F. Fogelman Soulié and G. Weisbuch, Random iterations of threshold networks and associative memory. *SIAM J. Comput.*, (1986)

[28] F. Fogelman Soulié and Y. Le Cun, Modèles connexionnistes de l'apprentissage. Apprentissage et Machine, *Intellectica*, 1, n°2.3. J. Quinqueton, J. Sallantin (Eds) Ass. Rech. cognitive (1987), 114–43

[29] K. Fukushima and S. Miyake, A self-organizing neural network with a function of associative memory: feedback-type cognitron, *Biol. Cyber.*, 28 (1978), 201–8

[30] S. Geman and D. Geman, Stochastic relaxation, Gibbs distributions, and the Bayesian restoration of images, *IEEE Trans. PAMI*, 6, 6 (1984), 721–41

[31] E. Goles Chacc, Comportement dynamique de réseaux d'automates, Thesis, Grenoble, (1985)

[32] E. Goles, this volume, Chapter 4

[33] T. N. E. Greville, Some applications of the pseudo inverse of a matrix, *SIAM Rev.*, 2 (1960), 15–22

[34] D. O. Hebb, *The Organization of Behavior*, Wiley, New York (1949)

[35] W. D. Hillis, New computer architectures and their relationship to physics or Why computer science is no good, *Int. J. Theor. Phys.* 21, 3/4 (1982), 255–62.

[36] W. D. Hillis, *The Connection Machine*, MIT Press, Cambridge, Mass. (1986)

[37] G. E. Hinton and T. Sejnowski, Optimal perceptual inference, *IEEE Conf. on Computer Vision and Pattern Recognition*, (1983), 448–53

[38] G. E. Hinton, Distributed representations. Working paper, CMU-CS-84-157 (1984)

[39] G. E. Hinton, private communication (1985)

[40] G. E. Hinton, Learning in parallel networks, *Byte* (April 1985), 265–73.

[41] G. E. Hinton, Learning to recognize shapes in a parallel network. In the proceedings of the Fyssen meeting on vision, Paris (March 1986)

[42] G. E. Hinton and J. A. Anderson (eds), *Parallel Models of Associative Memory*, Erlbaum, Hillsdale (1981)

[43] T. Hogg and B. A. Huberman, Understanding biological computation: reliable learning and recognition, *Proce. Natl. Acad. Sci. USA*, 8[(1984), 6871–5

[44] M. L. Honig and D. G. Messerschmitt, *Adaptive Filters, Structures; Algorithms and Applications*, Kluwer (1984)

[45] J. J. Hopfield, Neural networks and physical systems with emergent collective computational abilities, *Proc. Natl. Acad. Sci. USA*, 79 (1982), 2554–8

[46] S. Kirkpatrick, C. D. Gelatt and M. P. Vecchi, Optimization by simulated annealing, *Science*, 220, 4598 (1983), 671–80

[47] T. Kohonen and M. Ruohonen, Representation of associated data by matrix operators, *IEEE Trans. Comput.* (July 1973)

[48] T. Kohonen, An adaptive associative memory principle, *IEEE Trans. Comput.*, (April 1974)

[49] T. Kohonen, *Self-Organization and Associative Memory*, Springer Series in Information Sciences, vol. 8, Springer Verlag, Berlin, Heidelberg, New York (1984)

[50] Y. Le Cun, A learning scheme for asymmetric threshold network, in 'Cognitiva 85', CESTA-AFCET (ed.) (1985), 599–604 (in French)

[51] Y. Le Cun, Learning process in an asymmetric threshold network, in [12], 233–40

[52] W. A. Little, Existance of persistent states in the brain. *Math. Biosc.*, 19 (1974), 101–20

[53] J. L. McClelland and D. E. Rumelhart, Distributed memory and the representation of general and specific information, *J. Exp. Psychology*, 114, 2 (1985), 159–88

[54] W. S. McCulloch and W. Pitts, A logical calculus of the ideas immanent in nervous activity, *Bull. Math. Biophysics*, 5 (1943), 115–33

[55] D. Marr and T. Poggio, Cooperative computation of stereo disparity, *Science*, 194 (1976), 283–7

[56] M. Minsky and S. Papert, *Perceptrons, an Introduction to Computational Geometry*, MIT Press, Cambridge, Mass. (1969)

[57] S. Miyake and K. Fukushima, A neural network model for the mechanism of feature extraction, *Biol. Cybern.*, 50 (1984), 377–84

[58] K. Nakano, Associatron, a model of associative memory, *IEEE Trans. Syst. Man. & Cybern.*, SMC-2, 3 (July 1972)

[59] J. von Neumann, Theory of self reproducing utomata. A. W. Burks Ed., University of Illinois Press, Urbana (1966)

[60] N. J. Nilsson, *Learning Machines*, McGraw-Hill, New York (1965)

[61] P. Peretto, Collective properties of neural networks: a statistical physics approach, *Biol. Cybern.*, 50 (1984), 51–62

[62] T. Poggio, V. Torre and C. Koch, Computational vision and regularization theory, *Nature*, 317, 26 (1985), 314–19

[63] D. C. Plaut, S. J. Nolan and G. E. Hinton, Experiments on learning by back propagation. CMU-CS 86–126 (1986)

[64] L. D. Pyle, Generalized inverse computations using the gradient projection method. *J. Assoc. Comput. Mach.*, 11 (1964), 422–8

[65] C. R. Rao and S. K. Mitra, *Generalized Inverse of Matrices and its Applications*, Wiley, New York (1971)

[66] F. Robert, *Discrete Iterations, a metric study*, Springer Verlag, Berlin, Heidelberg, New York (1986)

[67] F. Robert, this volume, Chapter 1

[68] C. R. Rosenberg and T. J. Sejnowski, Practice on NETtalk a massively parallel network that learns to read aloud. *Connectionist Models: a summer school* (1986)

[69] F. Rosenblatt, *Principles of Neurodynamics*, Spartan, (1962)

[70] D. E. Rumelhart, J. L. MacClelland (eds), *Parallel and Distributed Processing: Explorations in the Microstructure of Cognition*, MIT Press, Cambridge, Mass. (1986)

[71] D. E. Rumelhart, G. E. Hinton and R. J. Williams, Learning internal representations by error propagation, in [70] vol 1, 318–62

[72] D. E. Rumelhart, J. L. MacClelland, On learning the past tenses of English verbs, in [70] vol 2, 216–71

[73] D. Sabbah, Computing with connections in visual recognition of Origami objects, *Cognitive Science*, 9 (1985), 25–50

[74] T. J. Sejnowski, and G. E. Hinton, Separating figure from ground with a Boltzmann machine. In *Vision, Brain and Cooperative Computation*, M. A. Arbib and A. R. Hanson (eds), MIT Press, Cambridge, Mass. (1985)

[75] T. J. Sejnowski, P. K. Kienker and G. E. Hinton, Learning symmetry groups with hidden units: beyond the perceptron, *Physica*, 22D, (1986)

[76] T. J. Sejnowski and C. H. Rosenberg, NETtalk: a parallel network that learns to read aloud, Johns Hopkins Technical report JHU/EECS-86/01, (1986)

[77] J. Sklansky and G. N. Wassel, *Pattern Classifiers and Trainable Machines*, Springer Verlag, Berlin, Heidelberg, New York (1981)

[78] G. S. Stiles and D. L. Denq, On the effect of noise on the Moore–Penrose generalized inverse associative memory, *IEEE Trans. PAMI*, 7, 3 (1985), 358–60

[79] D. L. Waltz and J. B. Pollack, Massively parallel parsing: a strongly interactive model of natural language interpretation, *Cognitive Science*, 9 (1985), 51–74

[80] G. Weisbuch and F. Fogelman Soulie, Scaling laws for the attractors of Hopfield networks, *J. Phys. Lett.*, 46 (1985), 623–30

[81] B. Widrow and M. E. Hoff, Adaptive switching circuits, *IRE WESCON Conv. Record, part 4*, (1960), 96–104

Systolic algorithms and architectures

8.1 Introduction

The traditional approach for high-performance computers pushes component speed and relies on elaborate engineering to minimise clock periods [28]. However, it seems difficult to obtain another order-of-magnitude speed-up in the near future by relying solely on component speed improvements, because we are close to various physical limits (such as the speed of light). The other approach is to use parallel systems: the concept of parallel processing is a departure from the trend of achieving increases in speed by performing single operations faster. Parallel processing achieves increases in speed by performing several operations in parallel [24]. Indeed, super-computers with vector and/or parallel computing facilities are becoming available to an increasing extent for scientific computation. Such parallel computing resources can be efficiently implemented with today's high-circuit-density VLSI technology, provided that the two following conditions are satisfied:

(i) The resource is a special-purpose computing device which is attached to a host architecture:

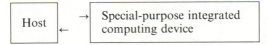

(ii) The application run on the resource is a computation-bound application, where the number of arithmetic operations is much greater than the number of input and output elements. Front-end processing in radar, sonar, vision, or robotics yields typical examples of computation-bound applications.

Systolic arrays, as defined by Kung and Leiserson [30], are a useful tool for designing special-purpose VLSI chips. A chip based on a systolic design consists essentially of a few types of very simple cells which are mesh-interconnected in a regular and modular way, and achieves high performance through extensive concurrent and pipeline use of the cells. The most usual interconnection geometries of systolic arrays are depicted in Fig. 8.1.

The name *systolic* given by Kung and Leiserson [30] is taken from the physiology of living beings. A systole is a contraction of the heart, by means of which blood is pumped to the different components of the organism. In a systolic system, the information (data and instructions) is rhythmically given into a structure of elementary processors (cells), to be processed and passed

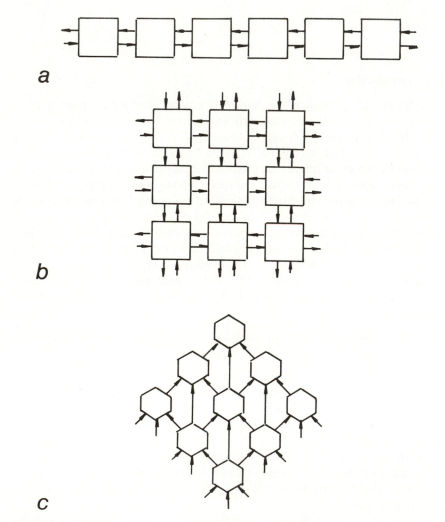

Figure 8.1 Examples of communication geometries: (a) linear array, (b) rectangular array, (c) hexagonal array

to a neighbour cell until a result reaches some border of the system communicating with the host [39].

To establish a connection between systolic architectures and cellular automata, we can say that systolic algorithms are algorithms which may be efficiently realised on systems that are representable as finite deterministic time-invariant synchronous cellular automata [39, 40]. Computational structures like systolic arrays have their foundations in the theory of cellular automata [8, 44] and iterative systems [19]. Some interesting concrete examples were disclosed in the early sixties [3, 9], however the technology was not yet sufficiently advanced to offer low complexity realisations. With the development of VLSI, the situation has changéd drastically.

Systolic arrays can now be realised as special-purpose integrated computing devices attached to a host architecture. Such regular networks of tightly coupled simple processors with limited storage have provided cost-effective high-throughput implementations of important algorithms in a variety of areas (signal and image processing, speech and pattern recognition, matrix computations). See [1, 5, 6, 12, 23, 26, 27, 31–4, 36, 41, 42, 52–4, 56, 57] among others.

Kung carefully explains in [25, 27] how such architectures should result in cost-effective and high-performance special-purpose systems, and he gives three main reasons to justify the advantages of the systolic model:

systolic systems are easy to implement because of their simple and regular design

they are easy to reconfigure because of their modularity

they permit multiple computations for each memory access, which speeds up the execution without increasing I/O requirements.

Heller [17] discusses the concept of a hardware library, where functional units are related to the host computer just as subroutines from a software library are to a production code. The original view of systolic arrays as external functional units attached to a host through various controllers fits into this model, and is illustrated in Fig. 8.2 [20, 32]. Heller also argues that systolic arrays can make good internal functional units, taking the role for simple matrix computations that vector pipelines now have for simple vector computations. This chapter describes some systolic arrays for the basic matrix algorithms that can be viewed both as external units to speed up the execution of pieces of program run on the host, and as the functional elements of a 'matricialisation' internal arithmetic unit.

The chapter is organised as follows. First some simple examples illustrate the basic concepts of the systolic model. We examine in Sections 8.2.1 and 8.2.2 a systolic algorithm for the one-dimensional convolution problem (nonrecursive and recursive). A systolic array for matrix–vector multiplication is introduced in Section 8.2.3. We deal with the solution of triangular linear systems in Sections 8.2.4. and 8.2.5. In Section 8.3.1, we consider a more difficult example: the triangularisation of a matrix on a two-dimensional systolic array, using either Gaussian elimination or Givens factorisation. Finally, a systolic array for the Jordan diagonalisation algorithm is depicted in Section 8.3.2.

These examples should give the reader a good insight of what type of

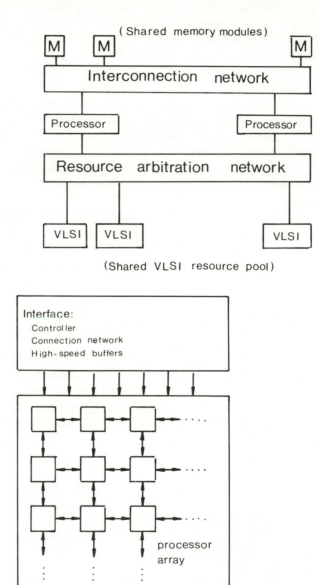

Figure 8.2 Hardware library

problems could be efficiently solved using systolic architectures. We come back to the advantages of the systolic model in Section 8.4.

The last section of the chapter is more theoretical. Indeed, Section 8.5 describes the methodology of Leiserson and Saxe [37] which aims to systematically derive systolic architectures from synchronous circuits.

8.2 Some simple systolic arrays

In this section, we consider some simple systolic arrays for problems arising from elementary linear algebra: one-dimensional convolution (nonrecursive and recursive), matrix-vector multiplication, solution of triangular systems of equations, triangular matrix inversion.

8.2.1 *Nonrecursive convolution*

Consider the following problem, called nonrecursive convolution or FIR filtering: given an input sequence x_1, x_2, x_3, ..., x_i, ..., compute for all $i \geq k$ the output

$$y_i = a_1 * x_i + a_2 * x_{i-1} + a_3 * x_{i-2} + \ldots + a_k * x_{i-k+1}$$

where the weights a_1, a_2, ..., a_k are fixed coefficients. A straightforward solution for the hardware implementation of the filter is depicted in Fig. 8.3 (with $k = 4$).

The operation of the filter is easy to understand. A new x_i is input every time-step. The k multiplications are performed simultaneously. Then we sum up the k partial results, using a k-input adder. A new y_i is computed every cycle time τ_k, but τ_k must be long enough to allow for the realisation of a multiplication followed by a k-input addition.

Another solution (a systolic solution !) is presented in Fig. 8.4. We still use $k = 4$ for the purpose of illustration.

The array is composed of k identical cells, whose operation is detailed in Fig. 8.5. The cells are called Inner Product Step (IPS) cells, since they are capable of performing a multiply-and-add.

There are two flows of data in the array. The x's move from left to right, reaching a new cell every time-step, without modification during their travel. The host delivers a new x_i every second time-step. The y's move at the same speed as the x's, in the opposite direction. They are computed through the successive accumulation of their k partial terms. A variable y_i is initially given the value 0; when it is output to the ost by the leftmost cell of the array, it has accumulated its final value

$$y_i = a_1 * x_i + a_2 * x_{i-1} + a_3 * x_{i-2} + a_4 * x_{i-3}$$

To see this, let t be the time-step where y_{i+2} is the next input of the array, with initial value $y_{i+2} = 0$ (see Fig. 8.6). The cells are numbered 1 to 4 from left

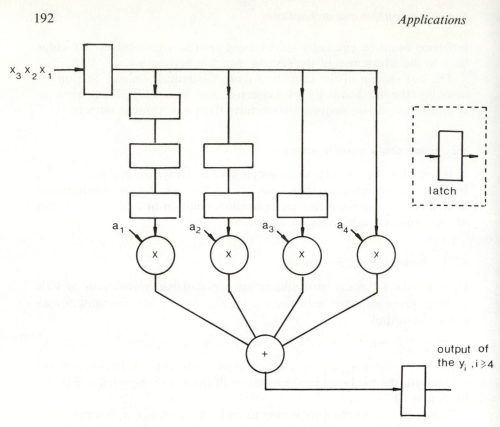

Figure 8.3 A first solution for nonrecursive convolution

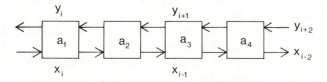

Figure 8.4 A systolic array for nonrecursive convolution

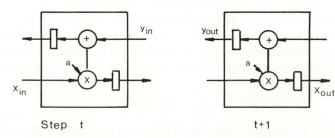

Figure 8.5 An IPS cell $y_{out}: = y_{in} + a*x_{in}$; $x_{out}: = x_{in}$

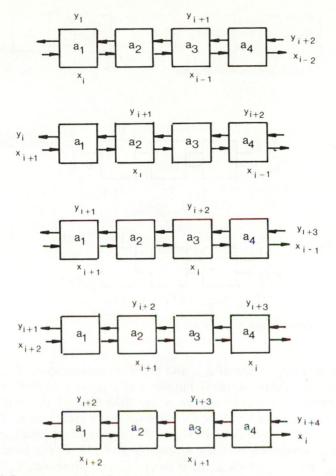

Figure 8.6 Some consecutive time-steps

to right. At time $t + 1$, the first accumulation $y_{i+2} := y_{i+2} + a_4*x_{i-1}$ takes place in cell 4. Then we have the accumulation $y_{i+2} := y_{i+2} + a_3*x_i$ at time $t + 2$ in cell 3, and the accumulation $y_{i+2} := y_{i+2} + a_2*x_{i+1}$ at time $t + 3$ in cell 2. The final value of y_i is obtained at time $t + 4$, with the last accumulation $y_{i+2} := y_{i+2} + a_1*x_{i+2}$ in cell 1.

We note that only the leftmost cell actually communicates with the host. The y input port of the rightmost cell does not exist, since the y's are initialised to 0. We show the communication between the array and the outside world in Fig. 8.7. Each x_i is input only once from the host, but it is reused many times in the array. This permits us to improve the performance without increasing the I/O requirements.

The computation of several y's is performed simultaneously: the array operates in a pipeline fashion. For instance, during the computation of y_{i+2},

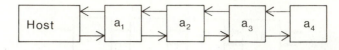

Figure 8.7 Communication with the host

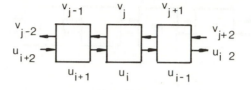

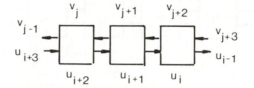

Figure 8.8 v_j meets u_i and u_{i+2} but not u_{i+1}

we finish the computation of y_{i+1} and begin the computation of y_{i+3}. After an initialisation delay, the array outputs a new y_i every second time-step.

This output rate could appear worse than that of the first solution depicted in Fig. 8.3, where a new y_i is output every cycle time τ_k. However, the cycle time τ_{syst} of the systolic array is that of a single cell, and corresponds to the time needed to perform a multiply-and-add. For k large enough, $\tau_k > 2\tau_{\text{syst}}$, and the systolic array is more efficient. Moreover, τ_{syst} does not depend on k, the length of the filter.

The major drawback of the array is that the cells are idle every second step. In fact, this is a general rule. Consider two flows of data (u_i) and (v_j) which move in opposite directions on a linear array: if two consecutive values are separated by a single time-step, then any variable in a given flow meets only one-half of the variables flowing in the opposite direction, as illustrated in Fig. 8.8.

Various solutions have been proposed to increase the efficiency of systolic arrays: consider a problem (P) solved on a systolic array where each cell is activated only every kth time-step. Kung and Leiserson [30] suggest we coalesce k adjacent processors into a single one. This reduces the area by a factor of k but does not improve the computation time. A second solution consists of interleaving the solution of k independent samples of (P), which does not speed up the execution of each sample of (P). A detailed discussion

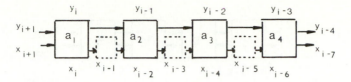

Figure 8.9 A systolic array where the x's and the y's move in the same direction

of this problem and various systolic designs of full efficiency can be found in [47, 49].

In fact, for the nonrecursive convolution problem, we can design a systolic array where a new output is computed every time-step [27]. There are k IPS cells in the array, whose internal registers are preloaded with the weighting coefficients. We let the x's and the y's move in the same direction, but the x's move twice as slowly as the y's. This is obtained by inserting additional delay cells on the x's path. This new systolic array is depicted in Fig. 8.9.

We shall not consider this array further, since it cannot be extended to handle the recursive convolution problem, unlike the array of Fig. 8.4. Kung [27] presents a comprehensive overview of systolic architectures for nonrecursive convolution.

8.2.2 Recursive convolution

We consider the following recurrent computation:

$$y_i = a_1 * y_{i-1} + a_2 * y_{i-2} + a_3 * y_{i-3} + \ldots + a_k * y_{i-k}$$

for $i \geq k + 1$; $y_1, y_2, \ldots, y_k$ are initial values. We simply add a delay cell to the array of Fig. 8.4 to deal with this computation, as shown in Fig. 8.10 (with $k = 4$). The y's are initialised to 0 and first move from right to left to accumulate their final value, just as before. Once they have reached the leftmost cell, they are reflected by the delay cell and move back in the opposite direction, playing the part of the x's in the nonrecursive problem. Hence they are not modified while moving back to the right.

Let t be the time-step represented in Fig. 8.10:

y_{i-1} and y_{i-2} have accumulated their final value and move rightwards without being modified

y_i performs the accumulation $y_i := y_i + a_1 * y_{i-1}$, which completes its computation

y_{i+1} performs the accumulation $y_{i+1} := y_{i+1} + a_3 * y_{i-2}$, which corresponds to its second partial term.

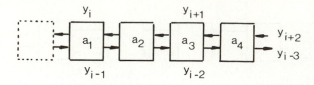

Figure 8.10 Recursive convolution array (time-step *t*)

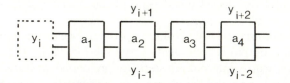

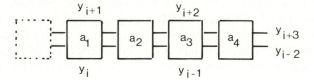

Figure 8.11 Recursive convolution array (time-steps *t* + 1 and *t* + 2)

At time *t* + 1 (see Fig. 8.11):

y_i is latched in the delay cell
y_{i+1} performs the accumulation $y_{i+1}: = y_{i+1} + a_3*y_{i-2}$, which corresponds to its third partial term.

At time *t* + 2 (see Fig. 8.11): y_i moves rightwards and is just in time to meet y_{i+1} in cell 1: thus y_{i+1} can complete its computation.

These simple considerations could be used for a formal proof of the correctness of the array, based on the invariance under translation.

After an initialisation delay, the array delivers a new y_i every second time-step, and the cycle time is independent of the order *k* of the recurrence.

The general recursive convolution problem, or IIR filtering, consists of computing

$$y_i = a_1*x_i + a_2*x_{i-1} + a_3*x_{i-2} + \ldots + a_k*x_{i-k+1}$$
$$+ r_1*y_{i-1} + r_2*y_{i-2} + r_3*y_{i-3} + \ldots + r_h*y_{i-h}$$

The array depicted in Fig. 8.12 solves this problem using *h* + *k* IPS cells and a delay cell. A new y_i is delivered every second time-step, and the cycle delay depends neither on the length *k* of the filter, nor on the order *h* of the recurrence.

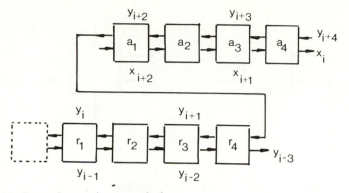

Figure 8.12 General recursive convolution

8.2.3 *Matrix–vector multiplication*

To compute the matrix–vector product $\mathbf{y} = \mathbf{Ax}$, where $\mathbf{A} = (a_{ij})$ is an $n \times n$ matrix and $\mathbf{x}$, $\mathbf{y}$ are n-tuples, the following recurrence can be used [36, 38]:

$$y_i^{(0)} = 0$$

$$y_i^{(k)} = y_i^{(k-1)} + a_{ik}{}^*x_k$$

$$y_i = y_i^{(n)}$$

Thus the only computations that are done using this recurrence are IPS computations. We use a linearly connected array of IPS cells which operate as indicated in Fig. 8.13. The only difference from the cells of Sections 8.2.1 and 8.2.2 is that the internal register of the cells is now dynamically updated through a vertical input/output port.

Figure 8.14 shows an example with $n = 4$, corresponding to the multiplication

$$\begin{bmatrix} y_1 \\ y_2 \\ y_3 \\ y_4 \end{bmatrix} = \begin{bmatrix} a_{11}\ a_{12}\ a_{13}\ a_{14} \\ a_{21}\ a_{22}\ a_{23}\ a_{24} \\ a_{31}\ a_{32}\ a_{33}\ a_{34} \\ a_{41}\ a_{42}\ a_{43}\ a_{44} \end{bmatrix} \begin{bmatrix} x_1 \\ x_2 \\ x_3 \\ x_4 \end{bmatrix}$$

The algorithm operates by pumping the data synchronously through the array as outlined in the figure:

The diagonals of the matrix $\mathbf{A}$ are fed in from the top, a new element being input to each cell every second step.
The components of the vector $\mathbf{x}$ are pumped in from the right, and move leftwards without modification.
The components of the vector y are fed through the left-hand end. Initially, the y_i's are set to 0, and in passing through the array, each y_i accumulates successively all of its terms $a_{i1}{}^*x_1, a_{i2}{}^*x_2, \ldots, a_{in}{}^*x_n$.

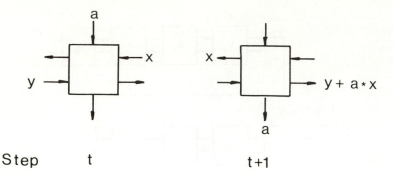

Step   t   t+1

Figure 8.13 IPS cell for matrix–vector multiplication

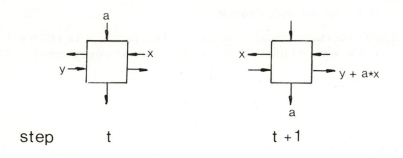

step   t   t +1

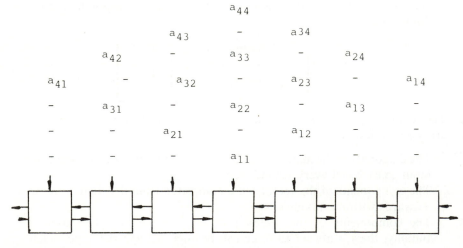

Figure 8.14 Matrix–vector multiplication, with $n = 4$

There are $2n - 1$ cells in the array. It is clear that after $2n - 1$ steps of computation, the y_i's start emerging from the right-hand end of the array, and that $2(n - 1)$ steps later, the last component y_n of **y** leaves the array. Thus the array takes $4n - 3$ steps to do the matrix–vector multiplication.

Note that, from the way in which the data are staggered, only half the processors are actually active at a given step of computation. The reason for interleaving a space between two consecutive components of the vectors **x** and **y** is exactly the same as in Section 8.2.1.

Let us define the efficiency as the ratio $e = T_{seq}/(p.T_p)$ of the (sequential) time T_{seq} needed to solve the problem with a single processor over the product of the number of cells p by the execution time T_p needed to solve the problem with p cells [20]. Since the processors are active every second step, we would expect the array to reach an efficiency $e = \frac{1}{2}$. However, the sequential number of steps is n^2, and we have $e = n^2/[(2n - 1)(4n - 3)]$, that is $e = \frac{1}{8}$ for large n. This is due to the initialisation of the array: for instance, no computation occurs during the first $n - 1$ steps, while x_1 and y_1 move in opposite direction to meet in the centre cell.

It is important to point out the impossibility of loading x_1 from the host directly into the centre cell at time $t = 1$, in order to avoid the initialisation delay. We assume that only the boundary cells can communicate with the host through their horizontal input and output ports. Internal cells communicate with the host only through their vertical input port. They receive the x_i's and the y_i's from their neighbours. Thus the centre cell has to wait for the outer cells to transmit x_1 and y_1 before it can start its computation. (In fact, there is no need to wait for y_1, since the y_i's are initialised to 0. We would have to wait for y_1 in the more general computation **y** $=$ **Ax** $+$ **d**. In any case, it is still necessary to wait for x_1.)

8.2.4 *Triangular systems of equations*

The triangular system solver described here is due to Kung and Leiserson [30], and solves the linear system of equations **Ax** $=$ **b**, where **A** is an $n \times n$ lower triangular matrix, and **b** $= (b_1, b_2, \ldots, b_n)^t$ is an n-vector (clearly any upper triangular system of equations can be reformulated as a lower triangular system). The solution $x = (x_1, x_2, \ldots, x_n)^t$ to this system can be computed using the following recurrence [36, 38]:

$$x_i^{(0)} = 0$$

$$x_i^{(k)} = x_i^{(k-1)} + a_{ik}{}^*x_k \text{ for } 1 \leq k \leq i - 1$$

$$x_i = (b_i - x_i^{(i-1)})/a_{ii}$$

The similarity of this defining recurrence to the previous one will lead us to a linearly connected systolic array similar to that for the matrix–vector

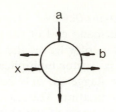
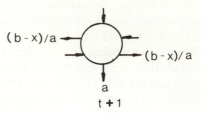

step   t   t + 1

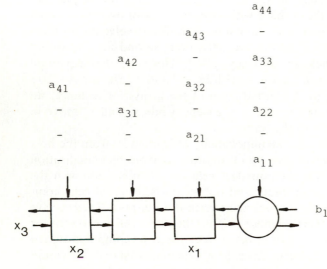

Figure 8.15 Triangular system of equations, with $n = 4$

multiplication. Let $n = 4$ as before; we have to solve the triangular system of
linear equations:

$$
\begin{bmatrix} a_{11} & 0 & 0 & 0 \\ a_{21} & a_{22} & 0 & 0 \\ a_{31} & a_{32} & a_{33} & 0 \\ a_{41} & a_{42} & a_{43} & a_{44} \end{bmatrix}
\begin{bmatrix} x_1 \\ x_2 \\ x_3 \\ x_4 \end{bmatrix}
=
\begin{bmatrix} b_1 \\ b_2 \\ b_3 \\ b_4 \end{bmatrix}
$$

The systolic array shown in Fig. 8.15 solves this system in the following
fashion: the coefficients given in the matrix **A** are pumped into the systolic
array from above, while the elements of the vector **b** are pumped into the
right-end processor from the right as shown.

There are two types of processor in the array: the first type (the square
processors) being IPS processors, and the second type (the circular processor
at the right-hand end) being the processor that computes the final value of
the solution elements in the vector **x**. The circular processor operates as
shown in Fig. 8.16.

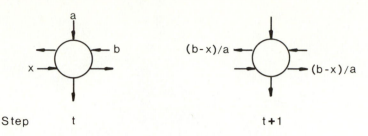

Step   t   t+1

Figure 8.16 Operation of the circular cell

When an element x_i initialised to 0 flows rightwards through the array, it accumulates its inner product terms

$$a_{i1}{}^*x_1 + a_{i2}{}^*x_2 + \ldots + a_{ii-1}{}^*x_{i-1}$$

At the special end processor, the final value of x_i is computed as

$$x_i := (b_i - x_i)/a_{ii}$$

Then x_i is fed back into the array and flows leftwards (without being modified), so that the following x_j, $j \geq i$, accumulate their inner product term $a_{ji}{}^*x_i$.

Let the computation start at time $t = 1$, when the circular processor computes $x_1 = b_1/a_{11}$. A new x_j is delivered by this processor every second step, hence the array solves the system in $2n - 1$ time-steps (there is no initialisation delay in this case).

8.2.5   *Application to triangular matrix inversion*

The above systolic algorithm for solving triangular systems can be extended to solve several systems of linear equations $Ax = b_j$, $1 \leq j \leq p$, where A is still a lower triangular matrix. We simply concatenate p copies of the previous array, and we let the column vector b_j flow through the jth copy, while the coefficients of the matrix A flow through the whole two-dimensional array (see Fig. 8.17 for an example with $n = p = 4$). By doing this, we are invariably solving the problem $AX = B$, where A is an $n \times n$ lower triangular matrix, B is an $n \times p$ matrix, and the solution X is an $n \times p$ matrix. Hence we are solving $X = A^{-1}B$. This computation requires $2n + p - 2$ time-steps on a two-dimensional array of $n*p$ cells.

Now by letting $p = n$ and $B = I_n$, the $n \times n$ identity matrix, we are solving the problem $X = A^{-1}$ within $3n - 2$ time-steps, using n^2 cells. In fact, we know *a priori* that X will be lower triangular too, and the cells below the diagonal can be suppressed since they only operate on zeros. This leads to the modified array of $n(n + 1)/2$ cells depicted in Fig. 8.18.

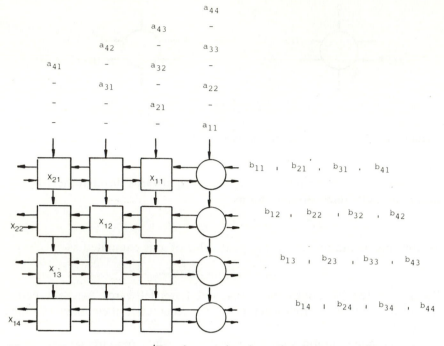

Figure 8.17 Computing $\mathbf{A}^{-1}\mathbf{B}$, where $\mathbf{A}$ is triangular $(n = p = 4)$

8.3 Systolic linear systems solvers

In this section, we deal with more complex examples. First we present the Gentleman–Kung triangularisation array in Section 8.3.1. Then we explain how to map the Gauss–Jordan diagonalisation algorithm onto a systolic architecture in Section 8.3.2.

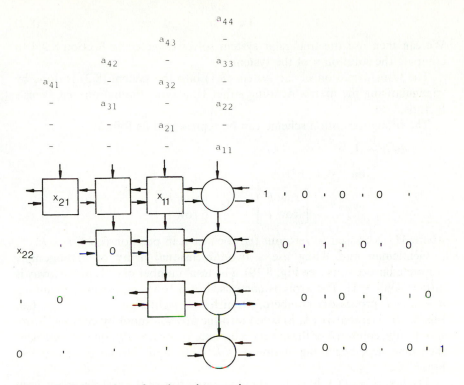

Figure 8.18 Inversion of a triangular matrix

8.3.1 *Matrix triangularisation*

Various systolic arrays have been proposed to triangularise a dense matrix: see [1, 4, 10, 16]. In this section we consider the array of Gentleman and Kung [16].

Let **A** be a dense $n \times n$ matrix and **b** an n-vector. To solve the system

$$\mathbf{Ax} = \mathbf{b} \tag{8.1}$$

we transform it into an equivalent triangular system

$$\mathbf{Tx} = \mathbf{b}' \tag{8.2}$$

We can then use the triangular system solver depicted in Section 8.2.4 to compute the solution **x** of the system.

The transformation of the system (8.1) into the system (8.2) is done by triangularising the matrix **A**, using either Gaussian elimination or Givens's factorisation.

The triangularisation scheme can be represented as follows:

do $k = 1, n$

    do $i = k + 1, n$

$$\begin{bmatrix} \text{row } k \\ \text{row } i \end{bmatrix} := M_{ik} \begin{bmatrix} \text{row } k \\ \text{row } i \end{bmatrix}$$

where M_{ik} is chosen to zero out the coefficient in position (i, k), $i > k$.

Gentleman and Kung use a two-dimensional array of orthogonally connected processors (see Fig. 8.19). The total number of cells in the array is $n[(n + 1)/2 + 1]$. The array is composed of n rows, each row k including $n + 2 - k$ processors numbered from left to right $P_{k,1}, \ldots, P_{k,n+2-k}$ (see Fig. 8.19). The matrix $(\mathbf{A}, \mathbf{b})$ is fed into the array column by column. More specifically, column k of the matrix is input to processor P_{1k}, one new element each time-step, beginning at time $t = k$. This input format is depicted in Fig. 8.19.

The first row of $(\mathbf{A}, \mathbf{b})$ is stored in the upper row of the systolic array, and, as any row numbered $i \geq 1$ is read by the array, it is combined with the first one in order to zero out the element a_{i1}; this combination corresponds to a transformation of the form

$$\begin{bmatrix} a_{11}, a_{12}, \ldots, a_{1n}, b_1 \\ a_{i1}, a_{i2}, \ldots, a_{in}, b_i \end{bmatrix} := \mathbf{M}_{i1} \begin{bmatrix} a_{11}, a_{12}, \ldots, a_{1n}, b_1 \\ a_{i1}, a_{i2}, \ldots, a_{in}, b_i \end{bmatrix}$$

The 2×2 matrix $\mathbf{M}_{i1}$ is computed by P_{11} at time $t = i$, and sent to the other cells of the first row; more precisely, the cell $P_{1,k}$ $(k \geq 2)$ performs the transformation

$$(a_{1k}, a_{ik})': = \mathbf{M}_{i1} \cdot (a_{1k}, a_{ik})'$$

at time $t = i + k - 1$.

In the case of Gaussian elimination, we have

$$\mathbf{M}_{i1} = \begin{bmatrix} 1 & 0 \\ -a_{i1}/a_{11} & 1 \end{bmatrix}$$

When the matrix **A** is not symmetric positive definite or diagonally dominant, we have to use an orthogonal factorisation for stability reasons. Pivoting is not possible; this would break down the regularity of the flow of the data in the array. We have to replace Gaussian elimination matrices by

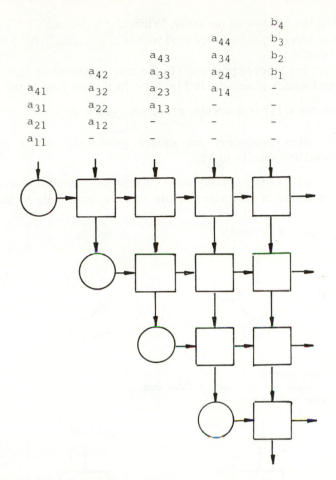

Figure 8.19 LU decomposition ($n = 4$)

Givens's rotation matrices. Hence,

$$\mathbf{M}_{i1} = \begin{bmatrix} \cos \theta & \sin \theta \\ -\sin \theta & \cos \theta \end{bmatrix}$$

where $\theta = \arctan(y/x)$. Let

$$(\mathbf{A}^{(k)}, \mathbf{b}^{(k)}) = \begin{bmatrix} a_{11}^{(k)} & \cdots & & a_{1n}^{(k)} & b_1^{(k)} \\ & a_{kk}^{(k)} & \cdots & a_{kn}^{(k)} & b_k^{(k)} \\ 0 & a_{nk}^{(k)} & \cdots & a_{nn}^{(k)} & b_n^{(k)} \end{bmatrix}$$

denote the matrix obtained from $(\mathbf{A}, \mathbf{b})$ after the elimination of the elements at positions (i, j) such that $i > j, j = 1, \ldots, k - 1$. Then

$$(a_{kk}^{(k)}, \ldots, a_{kn}^{(k)}, b_k^{(k)})$$

is stored in the kth row of the array. When $(a_{ik}^{(k)}, \ldots, a_{in}^{(k)}, b_i^{(k)})$, $i > k$, is read by this row of cells, it is combined with $(a_{kk}^{(k)}, \ldots, a_{kn}^{(k)}, b_k^{(k)})$ in order to set $a_{ik}^{(k)}$ to zero.

There are two types of processors in the array, respectively represented as circular and square processors in Fig. 8.19. In the kth row of the array:

the processor P_{k1} is a circular processor, which generates the matrices $\mathbf{M}_{ik}$, $i > k$.

all the other processors are square processors, which apply the transformation relative to $\mathbf{M}_{ik}$.

The operation of these two types of processors is detailed in Fig. 8.20 in the case of Gaussian elimination. Note that the only thing to modify for

Gaussian elimination

 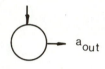

Step t Step $t + 1$

if init = true
then {store a_{in}}
  begin r: = a_{in}; init: = false; end
else {update a_{in}}
  a_{out}: = $- a_{in}/r$

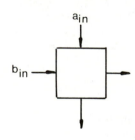 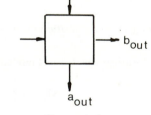

Step t Step $t + 1$

if init = true
then {store a_{in}}
  begin r: = a_{in}; init: = false; end
else {update a_{in}}
  begin a_{out}: = $a_{in} + b_{in}*r$;
         b_{out}: = b_{in};
  end

Figure 8.20 Operation of the processors for Gaussian elimination

Givens factorisation is the program of the processors: the global flow of the data remains the same in the array, regardless of the algorithm used to triangularise the matrix **A**. The program of the processors for Givens factorisation is detailed in Fig. 8.21. Of course, it would be very easy to

Givens factorisation

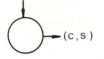

Step *t*   Step *t* + 1

if init = true
then {store a_{in}}
   begin r: = a_{in}; init: = false; **end**
else {generate a rotation}
   GENERATE (r, a_{in}, c, s)

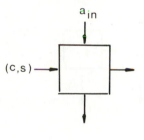
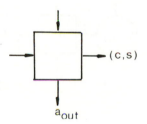

Step *t*   Step *t* + 1

if init = true
then {store a_{in}}
   begin r: = a_{in}; init: = false; **end**
else {apply the rotation (c, s)}
   begin APPLY (r, a_{in}, c, s); a_{out}: = a_{in}; **end**

Procedure GENERATE (x, y, c, s):
   {rotate (x, y) to annihilate y}
   if $y = 0$ **then begin** c: = 1; s: = 0 **end**
   else
      if $|y| \geq |x|$ **then begin** t: = x/y; s: = $1/(1 + t^2)^{1/2}$; c: = st **end**
      else begin t: = y/x; c: = $1/(1 + t^2)^{1/2}$; s: = ct **end**;
   x: = $cx + sy$; y: = nil;

Procedure APPLY (x, y, c, s):
   {apply rotation (c, s) to (x, y)}
   u: = x; v: = y; x: = $cu + sv$; y: = $-su + cv$;

Figure 8.21 Operation of the processors for Givens factorisation

combine both triangularisation schemes by implementing programmable cells.

The operation of a given processor in the array depends on whether it is the first data item it receives. As shown in Fig. 8.20 there is a control bit named 'init' (initialised to 'true') inside each processor which specifies the operation to be performed and the line along which the data is to be sent out. Figure 8.22 shows the content of the registers of the first row of the array, as well as its output format.

At the end, the matrix $(\mathbf{T}, \mathbf{b}')$ is stored in the array as follows:

$$t_{11}\ t_{12}\ t_{13}\ t_{14}\ b'_1$$
$$t_{22}\ t_{23}\ t_{24}\ b'_2$$
$$t_{33}\ t_{34}\ b'_3$$
$$t_{44}\ b'_4$$

There remains for us to unload the array; according to the general philosophy of the model, this must be done systolically, by propagating special boolean control instructions. Moreover, we would like to pipeline the

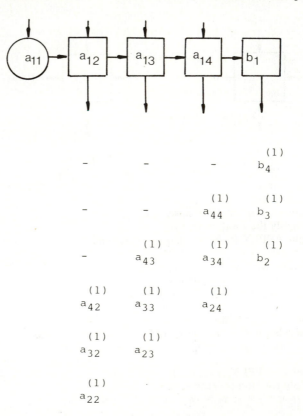

Figure 8.22 Operation of the first row of the array

unloading of the array with the computation phase, so that n additional steps are not necessary. Various schemes are possible. We outline a scheme where each processor sends the content of its register to the right when it has finished operating.

At time $t = 4$ in our example, P_{11} operates for the last time. It keeps idle for one step and then sends t_{11} rightwards. P_{12} finishes working at time $t = 5$. It transmits t_{11} to the right at $t = 6$, and then sends t_{12} at time $t = 7$. The process goes on, and t_{1i} is output by P_{15} at time $t = 9 + i$ ($1 \leq i \leq n + 1$, letting $b'_1 = t_{1,n+1}$). Similarly, we start unloading the second row at time $t = 8$. We see that the first rows of the array are unloaded, while the last rows still operate: this is the pipelining that we wanted. Indeed, in the general case, the last computation occurs in $P_{n,n+1}$ at time $3n - 1$, and $t_{i,i+k}$ ($1 \leq i \leq n, 0 \leq k \leq n+1-i$, letting $b'_i = t_{i,n+1}$) is output from the array at time $2n + i + k + 1$. The total computation time is $3n + 2$.

Finally, notice that we can simplify the unloading in the case of Gaussian elimination: once created, the $t_{i,i+k}$ are no longer modified, hence they can be sent downwards immediately. On the other hand, the $t_{i,i+k}$ are modified by each following rotation in the case of Givens factorisation.

8.3.2 *The Gauss–Jordan diagonalisation algorithm*

The systolic arrays presented in Sections 8.2.4 and 8.3.1 permit us to solve general dense systems of equations $Ax = b$: first triangularise the matrix (A, b) using the array of Section 8.3.1, then solve the triangular system with the array of Section 8.2.4.

Assume that there are p systems $Ax_i = b_i$ to solve, $1 \leq i \leq p$, and let B be the $n \times p$ matrix $B = (b_1, b_2, \ldots, b_p)$. To triangularise (A, B), simply extend the triangularisation array by adding $p - 1$ columns on the right, leading to a total number of cells $n[(n + 1)/2 + p]$. If $A = LT$, where L is either lower triangular with unit diagonal (Gaussian elimination) or orthogonal (Givens factorisation), and T is upper triangular, we obtain the matrix $(T, L^{-1}B)$ within $3n + p$ steps. There remains p triangular systems to be solved; this can be done within $2n + p$ steps using an array of pn cells (see Section 8.3). Hence we need a total number of $n^2/2 + 2pn + O(n)$ cells and of $5n + 2p$ time-steps to solve the p linear systems. However, this evaluation does not take into account the fact that the matrix $(T, L^{-1}B)$ has to be stored in the host and reordered by diagonals before entering the second systolic array.

We concentrate in this section of the design of a systolic array which directly computes the product $A^{-1}B$, where A is a dense (nonsingular) $n \times n$ matrix and B is a dense $n \times p$ matrix [47, 49]. The inverse of A is computed via the Gauss–Jordan diagonalisation algorithm. Just as in the LU decomposition of Section 8.4, special properties of A (symmetric positive definiteness or diagonal dominance) must be assumed to ensure good numerical accuracy.

Let $\mathbf{C}$ denote the $n \times (n + p)$ matrix $\mathbf{C} = (\mathbf{A}, \mathbf{B})$. We reduce $\mathbf{C}$ to the matrix $(\mathbf{I}, \mathbf{A}^{-1}\mathbf{B})$ by the Gauss–Jordan diagonalisation algorithm. The way to proceed is to premultiply $\mathbf{C}$ by n elementary $n \times n$ matrices $\mathbf{J}_1, \mathbf{J}_2, \ldots, \mathbf{J}_n$ in order to obtain after n steps

$$\mathbf{J}_n \ldots \mathbf{J}_2\mathbf{J}_1\mathbf{C} = (\mathbf{I}, \mathbf{A}^{-1}\mathbf{B})$$

Therefore, define $\mathbf{C}_k = \mathbf{J}_k \cdot \mathbf{C}_{k-1}$, starting with $\mathbf{C}_0 = \mathbf{C}$, and let $\mathbf{C}_k^0$ be the $n \times (n + p - k)$ matrix built up with the last $n + p - k$ columns of $\mathbf{C}_k$. We choose $\mathbf{J}_k$ so that the first k columns of $\mathbf{C}_k$ are those of $\mathbf{I}_n$, the identity matrix of order n. Thus $\mathbf{C}_k$ has the following structure:

$$\mathbf{C}_k = [\text{the first } k \text{ columns of } \mathbf{I}_n | \mathbf{C}_k^0]$$

In particular, $\mathbf{C}_n^0$ is the desired matrix $\mathbf{A}^{-1}\mathbf{B}$. The matrix $\mathbf{J}_k$ only differs from $\mathbf{I}_n$ by its kth column, which we denote as

$$[c_{1k}^{(k)} \ldots c_{nk}^{(k)}]^t$$

The coefficients of $\mathbf{C}_k^0$ are denoted $(c_{ij}^{(k)})$, $1 \leq i \leq n$, $k + 1 \leq j \leq n + p$. The values of the $c_{ij}^{(k)}$, $1 \leq i \leq n$, $k \leq j \leq n + p$, $1 \leq k \leq n$, are computed recursively by the following algorithm, starting from $c_{ij}^{(0)} = c_{ij}$:

```
for k: = 1 to n
begin
   {compute Jₖ}
   c_kk^(k): = 1/c_kk^(k-1)
   for i: = 1 to n, i ≠ k
      c_ik^(k): = - c_ik^(k-1)*c_kk^(k)
   {compute Cₖ⁰}
   for j: = k + 1 to n + p
   begin
      for i: = 1 to n, i ≠ k
         c_ij^(k): = c_ij^(k-1) + c_ik^(k)*c_kj^(k-1);
      c_kj^(k): = c_kk^(k)*c_kj^(k-1);
   end;
end;
```

Recall that $c_{ik}^{(k)}$ refers to the kth column of $\mathbf{J}_k$, while $c_{ij}^{(k)}$ with $k < j$ refers to the matrix $\mathbf{C}_k^0$.

We use a two-dimensional array of orthogonally connected processors (see Fig. 8.23). The array is composed of n rows, each row k including $n + 1$ processors numbered from left to right $P_{k,1}, \ldots, P_{k,n+1}$. The operation of each processor is detailed in Fig. 8.24. There are three types of processors:

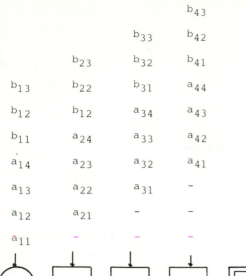

$$
\begin{array}{cccc}
 & & & b_{43} \\
 & & b_{33} & b_{42} \\
 & b_{23} & b_{32} & b_{41} \\
b_{13} & b_{22} & b_{31} & a_{44} \\
b_{12} & b_{12} & a_{34} & a_{43} \\
b_{11} & a_{24} & a_{33} & a_{42} \\
a_{14} & a_{23} & a_{32} & a_{41} \\
a_{13} & a_{22} & a_{31} & - \\
a_{12} & a_{21} & - & - \\
a_{11} & - & - & -
\end{array}
$$

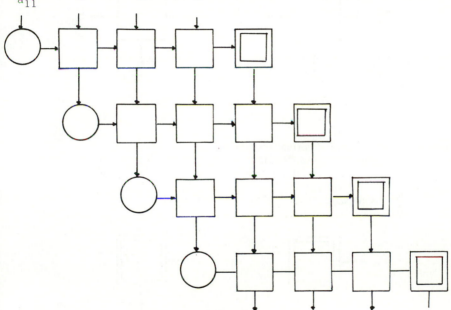

Figure 8.23 Computing $\mathbf{A}^{-1}\mathbf{B}$

type 1: circular processors compute the inverse of their first input data; afterwards they simply act as delay cells.

type 2: square processors first initialise their current register by storing after modification their first input data; then they act as IPS cells.

type 3: double-square processors actually operate as square processors, with the exception that their current register is not initialised in the same way (see Fig. 8.24)

Step *t*   Step *t* + 1

if init = true
then {perform division}
    begin a_{out}: = $1/a_{in}$; init: = false; **end**
else {transfer data} a_{out}: = a_{in}

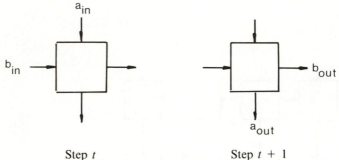

Step *t*   Step *t* + 1

if init = true
then {initialise current register}
    begin r: = $-a_{in}*b_{in}$; init: = false; b_{out}: = b_{in};
          a_{out}: = nil {no data is sent out downwards}; **end**
else {update a_{in}}
    begin a_{out}: = a_{in} + $r*b_{in}$; b_{out}: = b_{in}; **end**

Step *t*   Step *t* + 1

if init = true
then {initialise current register}
    begin r: = b_{in}; init: = false;
          a_{out}: = nil {no data is sent out downwards}; **end**
else {update b_{in}} a_{out}: = $r*b_{in}$

Figure 8.24 Operation of the processors for the Gauss–Jordan algorithm

In the kth row of the array, the leftmost processor $P_{k,1}$ is of type 1 and the rightmost processor $P_{k,n+1}$ is of type 3. All the other processors $P_{k,2}, \ldots, P_{k,n}$ are of type 2. As in Section 8.4.1, the operation of a given processor in the array depends on whether it is the first data item it receives. As shown in Fig. 8.24, there is a control bit named 'init' (initialised to 'true') inside each processor which specifies the operation to be performed and the line along which the data is to be sent out.

The matrix $\mathbf{A}$, followed by the matrix $\mathbf{B}$, is fed into the array row by row. More specifically, row k of the matrix $\mathbf{C}$ is input to processor P_{1k}, one new element each time-step, beginning at time $t = k$. This input format is depicted in Fig. 8.23.

The kth row of the array is devoted to the computation

$$\mathbf{C}_k{}^0 = \mathbf{J}_k \mathbf{C}_{k-1}{}^0$$

The leftmost processor $P_{k,1}$ transmits the input data arriving from the top to the right. Type 2 processors in the row modify and store the first data value arriving from the top and pass downwards all the following data after modification. Similarly the rightmost processor $P_{k,n+1}$ stores the first input data arriving from the left and passes downwards all the following data after modification. Thus after a row of $n + p$ input data flows through the whole array, its length is shortened by n to become a row of p output data. $\mathbf{C}_0 = \mathbf{C}_0{}^0 = \mathbf{C}$ is input to the first row of the array, $\mathbf{C}_1{}^0$ is input to the second, $\ldots$, and $\mathbf{C}_{n-1}{}^0$ is input to the nth row; finally the array outputs the matrix $\mathbf{C}_n{}^0 = \mathbf{A}^{-1}\mathbf{B}$.

It is important to note that the rows of the matrix are 'reordered' in some sense when moving through the array: the input of the processors $P_{k1}, \ldots, P_{k,n}$ in the kth row of the array are respectively the rows k, $k + 1, \ldots, n, 1, \ldots, k - 1$ of $\mathbf{C}_{k-1}{}^0$. The element $c_{kk}{}^{(k)}$ is computed according to $c_{kk}{}^{(k)} := 1/c_{kk}{}^{(k-1)}$ in processor $P_{k,1}$ and moves rightwards without modification until it is stored in the processor $P_{k,n+1}$. The nondiagonal coefficients $c_{ik}{}^{(k)}$, $i \neq k$, of the matrix $\mathbf{J}_k$ are computed and stored in the processors $P_{k,2}, \ldots, P_{k,n+1}$ (more precisely, $c_{ik}{}^{(k)}$ is stored in $P_{k,(i-k+1)\bmod n}$). After the processors are initialised, they perform the multiplication $\mathbf{C}_k{}^0 = \mathbf{J}_k \mathbf{C}_{k-1}{}^0$. The matrix $\mathbf{C}_k{}^0$ is output from the kth row of the array in row order, the leftmost row of the matrix being now row $(k + 1)$.

At step 1, processor P_{11} computes $c_{11}{}^{(1)} := 1/c_{11}{}^{(0)}$ and then operates as a one-step delay cell. At step 2, P_{12} initialises its current register with $c_{21}{}^{(1)} := -c_{21}{}^{(0)} * c_{11}{}^{(1)}$ and then operates as an IPS cell: indeed at step 3, it updates $c_{22}{}^{(0)}$ into

$$c_{22}{}^{(1)} := c_{22}{}^{(0)} + c_{21}{}^{(1)} * c_{12}{}^{(0)},$$

and so on with $c_{23}, c_{24}, \ldots$

Table 8.1 shows the operation of the first row of processors during phase 1

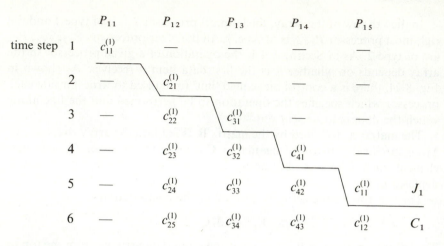

time step	P_{11}	P_{12}	P_{13}	P_{14}	P_{15}	
1	$c_{11}^{(1)}$	—	—	—	—	
2	—	$c_{21}^{(1)}$	—	—	—	
3	—	$c_{22}^{(1)}$	$c_{31}^{(1)}$	—	—	
4	—	$c_{23}^{(1)}$	$c_{32}^{(1)}$	$c_{41}^{(1)}$	—	
5	—	$c_{24}^{(1)}$	$c_{33}^{(1)}$	$c_{42}^{(1)}$	$c_{11}^{(1)}$	J_1
6	—	$c_{25}^{(1)}$	$c_{34}^{(1)}$	$c_{43}^{(1)}$	$c_{12}^{(1)}$	C_1

Table 8.1

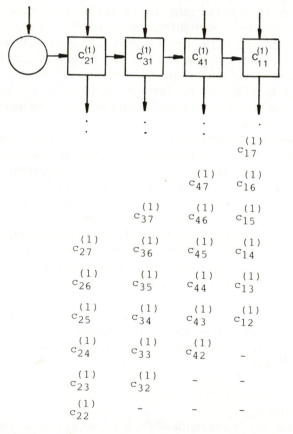

Figure 8.25 Computing $\mathbf{A}^{-1}\mathbf{B}$: operation of the first row of the array

(we choose $n = 4$ in this example). Figure 8.25 shows the content of the registers of the first row of the array, as well as its output format.

We can state the following: given a dense nonsingular $n \times n$ matrix **A** and a dense $n \times p$ matrix **B**, the orthogonal systolic array of $n(n + 1)$ processors can compute $\mathbf{A}^{-1}\mathbf{B}$ within $4n + p - 2$ time-steps. In particular, we can compute the inverse $\mathbf{A}^{-1}$ of a dense $n \times n$ matrix within $5n - 2$ steps, on a two-dimensional array of $n(n + 1)$ cells.

The computation of the product $\mathbf{A}^{-1}\mathbf{B}$ could also be achieved by using several of the previous existing arrays, together with an array for matrix–matrix multiplication. It is the approach of [21] to use such building blocks for: (i) computing the **LU** decomposition of **A**; (ii) inverting the triangular matrices **L** and **U**; (iii) computing the product $\mathbf{U}^{-1}\mathbf{L}^{-1} = \mathbf{A}^{-1}$; and (iv) computing the product $\mathbf{B}\mathbf{A}^{-1}$: however the price to pay is the extra delay needed to store partial results in the host, and most often to rearrange them in order to get the right input format for processing by the next building block (see the discussion in [40] for the problem of computing the product of three matrices). On the other hand, note the superiority of our 'one-array' approach for hardware implementation.

8.4 Why systolic architectures?

As the various examples of the previous sections should have demonstrated, the systolic model is a general methodology for making effective use of very large numbers of circuits in parallel. We come back in this section to the advantages of systolic architectures. So far, we have not given a precise definition of the word 'systolic'. Let us say until Section 8.5 that a systolic array is a regular synchronous array of simple and locally interconnected processors. This section is based on [13, 27], and its heading is the title of [27], a reference paper on systolic architectures.

8.4.1 *Concurrency and communication*

As already mentioned in the examples, several computations are performed by each data item inside a systolic array: once a data item is brought out from the external memory into the array, it passes from cell to cell and can be reused a number of times. This speeds up the execution without increasing I/O requirements, leading to the suppression of the 'I/O bottleneck' of traditional von Neumann machines (see Fig. 8.26, which is based on [27]).

An important restriction should be mentioned, however: although only the boundary cells communicate with the outside world, there is still a limitation concerning the I/O: this limitation lies in the number of available input/output pins of a systolic chip, and is especially true regarding two-dimensional arrays. Kung [27] emphasises the need of balancing computation with I/O.

Instead of

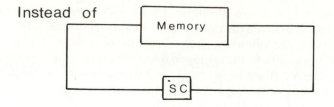

we have

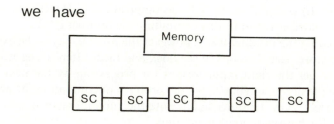

SC = systolic cell

Figure 8.26 Basic principle of a systolic system

8.4.2 *Simplicity and regularity*

Cost effectiveness is a major concern in designing special-purpose VLSI architectures: their cost must be low enough to justify their limited applicability [20]. Systolic arrays are composed of a few types of cells that are used repetitively with extremely simple interfaces (the latches): this is a first advantage. As a second advantage, the local and regular interconnection of the cells greatly facilitates the topological implementation, since the data path matches the geometric pattern of the array. Moreover, the control is reduced to a global clock distributed to all the cells. In the simplest case, that of one-dimensional linear arrays, any systolic algorithm can be straight forwardly mapped onto a silicon layout.

Let us return to the first convolution design depicted in Fig. 8.3. The x's are propagated along a global data bus, and duplicated to feed the four multipliers. The design of a filter with 5 (or more!) weights from the one with 4 weights is not easy, whereas with the systolic solution of Fig. 8.4 there is no problem in designing a filter of any length. More generally, systolic architectures enjoy a total modularity – due to the local interconnections and the elimination of global broadcasting – and a reduced fan-out (i.e. the maximum number of copies of the same variable created inside the array). As a result, they are easily reconfigurable.

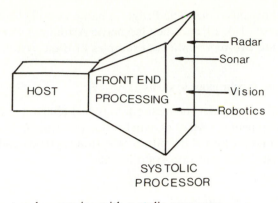

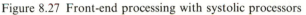

Figure 8.27 Front-end processing with systolic processors

Finally, we can make use of all these nice properties of simplicity, regularity and modularity to make our systolic arrays fault-tolerant. The key idea is to add some switches to all the cells, so that faulty cells can be bypassed. We refer to [29] for the design of fault-tolerant systolic systems.

8.4.3 *Systolic architectures are computation–intensive*

VLSI processing structures are suitable for implementing computation-bound algorithms rather than I/O-bound computations [20]. In a computation-bound algorithm, the number of computing operations is larger than the total number of input and output elements. Otherwise, the problem is I/O-bound. For example, the multiplication of two matrices of order n is a computation-bound problem, since there are n^3 multiply-and-adds and only $3n^2$ I/O operations. On the other hand, the addition of two matrices is I/O-bound, since there are n^2 additions and $3n^2$ I/O operations.

Computation-bound problems occur in front-end processing (see Fig. 8.27, based on [28]). For example, two dominating aspects in signal and image processing requirements are enormous throughput rates and huge amounts of data and memory [32]. Fortunately, most of the algorithms to be executed on the data are highly regular, since they fall into the classes of convolution or elementary numerical linear algebra algorithms. Most of the time, special-purpose systolic architectures will ensure a real-time performance solution.

A list of problems amenable to systolic solutions is given in [13]. These problems include all the fundamental signal and image processing algorithms, and the usual matrix arithmetic algorithms. It is worth pointing out that several nonnumerical applications (such as data structures and graph problems) can be found in the list.

Many versions of systolic processors are being designed by universities and industrial organisations. The Programmable Systolic Chip of Carnegie-Mellon University [13, 14] and the Geometric Arithmetic Parallel Processor of NCR [43] are two representative examples of these systolic designs.

8.5   Systolising synchronous systems

In this section, we consider the methodology of Leiserson and Saxe [37] for optimising the pipelining delay of synchronous systems by retiming techniques. We shall give a formal definition of systolic systems and show how to convert nonsystolic designs into systolic ones.

8.5.1   *A graph-theoretical model*

A circuit is formally defined as a directed graph $G = (\mathbf{V}, \mathbf{U})$. The set of vertices $\mathbf{V} = \{v_{\text{host}}, v_1, v_2, \ldots, v_n\}$ corresponds to the functional elements (combinational logic) of the circuit. We distinguish a special vertex v_{host} to represent the host processor, which is the system's only interface to the external world.

Each vertex v of G has a nonnegative propagation delay $d(v)$ which is the cycle time of the functional element that it represents. Each arc $e = (v, v')$ of $\mathbf{U}$ models a connection between the functional elements v and v'. It is weighted by an integer $w(e)$ which is equal to the number of latches along the interconnection. See Fig. 8.28 for an example.

If the weight of an edge happens to be zero, no latch impedes the propagation of a signal along the corresponding interconnection. For example, this is the case for the edge between v_2 and v_3 in the figure. If the weight of an elementary circuit in the graph happens to be zero (we take the weight of a path in the graph as the sum of the weights of its edges), some combinational rippling is fed back onto itself, and the value of the signal propagated along the circuit becomes uncertain. Hence we restrict ourselves

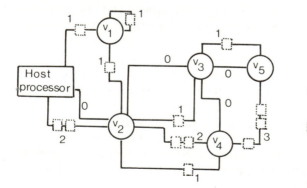
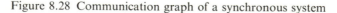

Figure 8.28 Communication graph of a synchronous system

to synchronous systems, defined as systems where every elementary circuit in the graph has nonzero weight [37].

The global cycle time of a synchronous system which operates in a pipeline fashion is equal to the longest path of combinational rippling, since it is the minimum period of the system clock that can be achieved. For example, assuming that all the vertices in the figure have the same propagation delay $d(v_i) = 1$ for $1 \leq i \leq 5$, the longest path is from the host to v_4 through v_2 and v_3, and its weight is 3. This means that the host can deliver a new input to the vertex v_2 every third unit of time.

We define a systolic system to be a system where each arc in the graph has a nonzero weight [37]. Such systems are obviously synchronous, and they exhibit no combinational rippling. As a consequence, they can operate at the rate of their slowest functional element: the cycle time of a systolic system is equal to max $\{d(v); v \in \mathbf{V}\}$. The example of Fig. 8.28 is synchronous but not systolic. If we could modify the repartition of the latches along the arcs so as to obtain an equivalent systolic system, we would obtain a pipeline cycle time of only one unit of time. The purpose of the methodology to be described in Section 8.5.3 is to transform synchronous systems into systolic ones, so as to reduce the period of the system clock. Before that, we return to the example of the one-dimensional nonrecursive convolution.

8.5.2 *The convolution example revisited*

We consider again the problem of computing

$$y_i = a_1 \cdot x_i + a_2 \cdot x_{i-1} + \ldots + a_k \cdot x_{i-k+1}$$

as introduced in Section 8.2.1. Consider the solution depicted in Fig. 8.3, and replace the k-input adder by $(k - 1)$ consecutive 2-input adders. We obtain the circuit of Fig. 8.29, for $k = 4$. With the graph formalism, we are led to the representation of Fig. 8.30.

Let ∂_{mult} denote the propagation delay of a multiplier and ∂_{add} that of an adder. In other words, considering Fig. 8.29, $d(v_i) = \partial_{\text{mult}}$ for $1 \leq i \leq 4$ and $d(v_i) = \partial_{\text{add}}$ for $5 \leq i \leq 7$. The longest path of combinational rippling has a

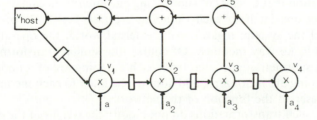

Figure 8.29 The solution of Fig. 8.3 with $(k - 1)$ 2-input adders

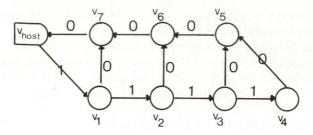

Figure 8.30 Representing the network of Fig. 8.29.

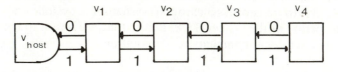

Figure 8.31 Coalescing vertices of the previous solution

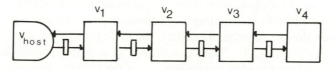

Figure 8.32 Representing the network of Fig. 8.31

length $\partial_{mult} + (k - 1) \cdot \partial_{add}$, and this corresponds to the cycle time of the system.

If we choose to coalesce a multiplier and an adder in a single vertex, we obtain the solution depicted in Figs. 8.31 and 8.32. Each vertex has a propagation delay equal to $\partial_{mult} + \partial_{add}$, and the longest path has now a length $k \cdot (\partial_{mult} + \partial_{add})$. Again, this is the cycle time of the system. We will illustrate the methodology with these two examples.

8.5.3 *A methodology for retiming*

Let v be a vertex whose every incoming arc has a nonzero weight. The transformation that consists of subtracting one latch from each incoming arc and adding one latch to each outgoing arc (see Fig. 8.33) preserves the function of the system: regarding the external world, the operation of the vertex v is in no way modified. Of course, the similar transformation that consists of subtracting one latch from each outgoing arc of a node v' whose every outgoing arc is nonzero, and adding one latch to each incoming arc of v', also preserves the function of the network.

Clearly, such transformations do not modify the weight of the elementary circuits of the network. As a consequence, a necessary condition for the

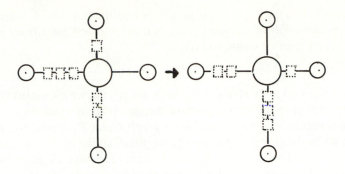

Figure 8.33 A transformation of the graph

conversion of the original system into a systolic one is the following: for every elementary circuit of the original graph, the number of latches is greater than or equal to the number of arcs. Remember that in a systolic system, each arc has a nonzero weight, so that each elementary circuit has a weight greater than or equal to the number of its arcs.

Another formulation of the previous condition is the following: for converting a system represented by a graph G into a systolic system, it is necessary that there exists no elementary circuit in the graph $G - 1$ whose weight is negative. Here $G - 1$ denotes the graph obtained by subtracting one latch on every arc of G. The weights of the arcs of $G - 1$ are taken in $\mathbb{Z}$.

Leiserson and Saxe have shown that this necessary condition is also sufficient. Indeed, assume there exists no negative elementary circuit in $G - 1$. We shall use the transformations of Fig. 8.33 to convert G into a systolic graph G'. We substract $r(i)$ latches from each outgoing arc of the vertex v_i, and we add $r(i)$ latches to each incoming arc of v_i, where $r(i) \in \mathbb{Z}$ remains to be defined. Thus the weight w_{ij} of an arc from v_i to v_j is changed into $w_{ij} + r(j) + r(i)$:

Graph G: Graph G':

$$v_i \xrightarrow{\;\;w_{ij}\;\;} v_j \quad\Rightarrow\quad v_i \xrightarrow{\;\;w_{ij} + r(j) - r(i)\;\;} v_j$$

We let $r(i)$ be the length of the shortest path from the vertex v_i to the host v_{host} in the graph G-1:

Graph G-1:

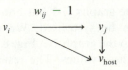

By definition of $r(i)$, the length of the path from v_i to the host which is composed of the concatenation of the arc (v_i, v_j) and the shortest path from v_j to the host is greater than or equal to $r(i)$:

$$r(i) \leq w_{ij} - 1 + r(j)$$

Therefore, $w_{ij} + r(j) - r(i) \leq 1$ for each arc (v_i, v_j), which means that each arc in the new graph G' has a positive weight: G' is systolic.

Given a network R represented by a graph G, let R_k denote the network represented by the graph kG, where the weight of all the arcs in the graph G is multiplied by k. The network R_k has the same function as the network R, but its speed has been divided by a factor of k.

We are led to the following methodology for designing systolic systems:

Step 1 define a simple network R with combinational rippling
Step 2 determine the smallest integer k such that the network R_k can be converted into a systolic network
Step 3 systolise R_k, using the previous transformations

Leiserson and Saxe give algorithms for performing efficiently steps 2 and 3. For the sake of simplicity, we only consider their general methodology on the two examples of the Section 8.5.2.

Consider the network of Figs. 8.29 and 8.30, which are reproduced in Fig. 8.34(a) for the sake of convenience. For the second step of the methodology, we want to determine the smallest k such that kG can be converted into a systolic graph. We check easily that there exist negative circuits in the graphs G-1 and $2G$-1, but none in $3G$-1 (see Fig. 8.34(b)). Hence, the graph $3G$ can be systolised.

Scanning out the graph $3G$-1, we can compute the value of $r(i)$ for $1 \leq i \leq 7$. We obtain easily: $r(1) = -2$, $r(2) = -3$, $r(3) = r(4) = -4$, $r(5) = -3$, $r(6) = -2$, and $r(7) = -1$. Coming back to the graph $3G$, Fig. 8.34(c), we can modify the re-partition of the latches along the arcs according to these values. The resulting graph G' is shown in Fig. 8.34(d). The corresponding systolic network is illustrated in Fig. 8.34(e). Since it is a 3-slowed version of the original network, its cycle time is equal to $3 * \max \{\partial_{\text{mult}}, \partial_{\text{add}}\}$. We give another representation of the network in Fig. 8.34(f), where we use the same convention as in Sections 8.2 and 8.3: we assume that synchronisation barriers are included in each functional element. In other words, a latch has been subtracted from each outgoing arc of the graph to be included in the corresponding functional element.

Consider now the network of Figs. 8.31 and 8.32, which we reproduce in Fig. 8.35(a). For the second step of the methodology, we check that there exist negative circuits in the graphs G-1 but none in $2G$-1 (see Fig. 8.35(b)). Hence, the graph $2G$ can be systolised. We obtain that $r(i) = -i$ for $1 \leq i \leq 4$. Coming back to the graph $2G$, Fig. 8.35(c), we modify the weight of the arcs according to these values. The resulting graph G' is shown in

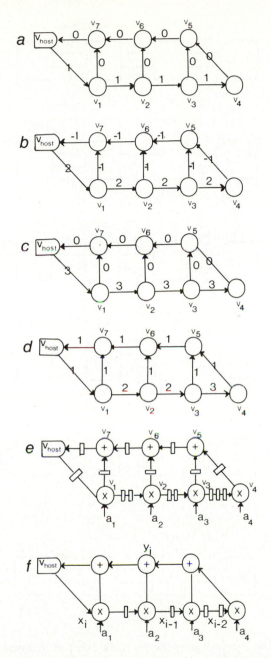

Figure 8.34 (a) The original graph G; (b) there exists no negative circuit in $3G$-1; (c) the graph $3G$; (d) the systolic graph G'; (e) the network represented by G'; (f) using the same convention as in the Section 8.2 and 8.3: latches are included in every functional unit.

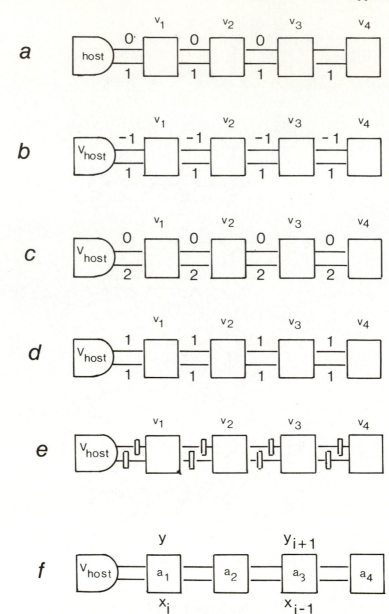

Figure 8.35 (a) the original graph G; (b) there exists no negative circuit in $2G$-1; (c) the graph $2G$; (d) the systolic graph G'; (e) the network represented by G'; (f) using the same convention as in the Sections 8.2 and 8.3: latches are included in every functional element.

Fig. 8.35(d), and the corresponding systolic network is illustrated in Fig. 8.35(e). Since it is a 2-slowed version of the original network, its cycle time is equal to $2 * (\partial_{mult} + \partial_{add})$. We give another representation of the network in Fig. 8.35(f), with the same convention as for Fig. 8.34(f).

Comparing the cycle time of the two original networks, we check that the first is always superior to the second. But when comparing the two systolic versions, the result depends on the ratio $\rho = \partial_{mult}/\partial_{add}$.

Indeed, we have to compare

$$\tau_1 = 3 * \max\{\partial_{mult}, \partial_{add}\} \text{ and } \tau_2 = 2 * (\partial_{mult} + \partial_{add})$$

Assuming that $\partial_{add} \leq \partial_{mult}$, we have $\tau_1 \leq \tau_2$ if and only if $\rho \leq 2$. This result would be of importance for the hardware implementation of a systolic convolution array.

8.6 Conclusion

Many systolic arrays have been introduced in the literature to solve particular instances of matrix algebra and digital signal processing problems. We have surveyed some of them in Section 8.2 and 8.3: nonrecursive and recursive convolution, matrix–vector multiplication, triangular system solution, triangular matrix inversion, **LU** or Givens triangularisation, computation of the product $\mathbf{A}^{-1}\mathbf{B}$, where **A** and **B** are general dense matrices.

Using massive parallelism and pipelining, the systolic array concept allows a system implementor to design extremely efficient machines for specific computations. The aim of this chapter was to support this claim with some examples and a brief presentation of the advantages of the systolic model. I hope that I have convinced the reader!

We have examined the methodology of Leiserson and Saxe in Section 8.5. The following chapter will describe systematic and automatic design of systolic arrays based upon uniform recurrence equations. See also the paper of Lam and Mostow [35], and the references therein, for a description of other methodologies towards the systolisation of special-purpose architectures.

We conclude this chapter with the following citation taken from the PhD thesis of C. E. Leiserson [36]:

Systolic systems are an attempt to capture the concepts of parallelism, pipelining and interconnection structures in a unified framework of mathematics and engineering. They embody engineering techniques such as multiprocessing and pipelining together with the more theoretical ideas of cellular automata and algorithms, and therefore are an excellent subject of investigation from a combined standpoint.

8.7 **References**

[1] H. M. Ahmed, J. M. Delosme and M. Morf, Highly concurrent computing structures for matrix arithmetic and signal processing, *Computer*, 15 (1982), 65–82

[2] F. Andre, P. Frison and P. Quinton, Algorithmes systoliques: de la théorie à la pratique, RR 214, INRIA ,Rocquencourt, Mai 1983

[3] A. J. Atrubin, A one-dimensional real time iterative multiplier, *IEEE Trans. Comput.*, 14, 6 (1965), 394–9

[4] A. Bojanczyk, R. P. Brent and H. T. Kung, Numerically stable solution of dense systems of linear equations using mesh-connected processors, Technical Report, Carnegie-Mellon University, Pittsburgh, Penn. (1981)

[5] R. P. Brent, H. T. Kung and F. T. Luk, Some linear-time algorithms for systolic arrays, Technical Report TR CS 82-541, Cornell University, Ithaca, NY (1982)

[6] P. R. Capello and K. Steiglitz, Completely pipelined architectures for digital signal processing, *IEEE Trans. ASSP*, 31, 4 (1983), 1016–23

[7] H. Y. H. Chuang and G. He, A versatile systolic array for matrix computations, *Proc. 12th Symposium on Computer Architecture, June 17–19, 1985*

[8] E. F. Codd, *Cellular Automata*, Academic Press, New York (1968)

[9] S. N. Cole, Real-time computation by *n*-dimensional iterative arrays of finite-state machines, *IEEE Trans. Comput.*, 18, 4 (1969), 349–65

[10] M. Cosnard and Y. Robert, Systolic Givens factorization of dense rectangular matrices, Research Report TIM3/IMAG Grenoble 590 (1986)

[11] M. Cosnard, Y. Robert and M. Tchuente, Matching parallel algorithms with architectures: a case study, *IFIP Conference on Highly Parallel Architectures, Nice, France, May 25–27, 1986*

[12] R. A. Evans, D. Wood, J. V. McCanny, J. B. McWhirter and A. P. H. McCabe, A CMos implementation of a systolic multi-bit convolver chip, *Proc. VLSI 83, Trondheim, Norway*, F. Anceau *et al.* (eds), North-Holland, Amsterdam (1983)

[13] A. L. Fisher and H. T. Kung, Special-purpose VLSI architectures: general discussions and a case study, in [34], 153–69

[14] A. L. Fisher, H. T. Kung and K. Sarocky, Experience with the CMU program-mable systolic chip, *Proc SPIE Symp., vol. 495, Real-Time Signal Processing VII* (1984)

[15] M. J. Foster and H. T. Kung, The design of special-purpose VLSI chips, *IEEE Computer*, 13, 1 (1980), 26–40

[16] W. M. Gentleman and H. T. Kung, Matrix triangularisation by systolic arrays, *Proc SPIE Symp, vol 298, Real-time Signal Processing IV* (1981), 19–26

[17] D. Heller, Partitioning big matrices for small systolic arrays, in [34], 185–99

[18] D. Heller and I. Ipsen, Systolic networks for orthogonal equivalence transformations and their applications, *Proc. 1982 Conf. Advanced Research in VLSI, MIT, Boston* (1982), 113–22

[19] F. C. Hennie, *Iterative Arrays of Logical Circuits*, MIT Press, Cambridge, Mass. (1961)

[20] K. Hwang and F. Briggs, *Parallel Processing and Computer Architecture*, MacGraw-Hill, New York (1984)

[21] K. Hwang and Y. H. Cheng, Partitioned matrix algorithm for VLSI arithmetic systems, *IEEE Trans. Comput.*, 31 (1982), 1215–24

[22] M. R. Kramer and J. Van Leeuwen, Systolic computation and VLSI, Foundations of Computer Science IV, J. W. DeBakker and J. Van Leeuwen (eds), (1983), 75–103

[23] A. V. Kulkarni and D. W. L. Yen, Systolic processing and an implementation for signal and image processing, *IEEE Trans. Comput.*, 31, 10 (1982), 1000–9

[24] S. P. Kumar, Parallel algorithms for solving linear equations on MIMD computers, Thesis, Washington State University, Pullman, Wash. (1982)

[25] H. T. Kung, The structure of parallel algorithms, *Advances in Computers*, 19 (1980), 65–112

[26] H. T. Kung, Special-purpose devices for signal and image processing, *Proc SPIE Symp, vol 241, Real-time Signal Processing III* (1980), 76–84

[27] H. T. Kung, Why systolic architectures? *Computer*, 15, 1 (1982), 37–46

[28] H. T. Kung, Programmable systolic chip, *Proc. NATO Advanced study institute on Microarchitecture of VLSI computers, Sogesta, Urbino, Italy, July 9–20, 1984*

[29] H. T. Kung and M. S. Lam, Fault-tolerance and two-level pipelining in VLSI systolic arrays, *J. Parallel & Distributed Computing*, 1, 1 (1984), 32–63

[30] H. T. Kung and C. E. Leiserson, Systolic arrays for (VLSI), *Proc. of the Symposium on Sparse Matrices Computations*, I. S. Duff *et al.* (eds), Knoxville, Tenn. (1978), 256–82

[31] S. Y. Kung, On supercomputing with systolic/wavefront array processors, *Proc. IEEE*, 72 (1984), 867–84

[32] S. Y. Kung, VLSI array processors, *IEEE ASSP Magazine*, 2, 3 (1985), 4–22

[33] S. Y. Kung and J. Annevelink, VLSI design for massively parallel signal processors, *Microprocessors and Microsystems* (Special issue on signal processing devices) 7, 4 (1983), 461–8

[34] S. Y. Kung, H. J. Whitehouse and T. Kailath (eds), *VLSI and Modern Signal Processing*, Prentice-Hall, Englewood Cliffs, NJ (1985)

[35] M. Lam and J. Mostow, A transformational model of VLSI systolic design, *Computer* 18, 2 (1985), 42–52

[36] C. E. Leiserson, Area-efficient VLSI computation, Thesis, Carnegie-Mellon University, Pittsburgh, Penn. (1981)

[37] C. E. Leiserson and J. B. Saxe, Optimizing synchronous systems, *Proc. of the 22th Annual Symposium on Foundations of Computer Science, October 1981*, 23–36

[38] F. Moller, A survey of systolic systems for solving the Algebraic Path Problem, Report CS-85-22, University of Waterloo, Ontario, Canada (1985)

[39] C. Moraga, Systolic algorithms, Technical Report, Computer Science Department, University of Dortmund, FRG (1984)

[40] C. Moraga, On a case of symbiosis between systolic arrays, *Integration, the VLSI journal*, 2 (1984), 243–53

[41] J. G. Nash and S. Hansen, Modified Faddeev algorithm for matrix manipulation, *Proc SPIE Symp., vol. 495, Real-Time Signal Processing VII* (1984)

[42] J. G. Nash, S. Hansen, and G. R. Nudd, VLSI processor arrays for matrix manipulation, *VLSI Systems and Computations*, Rockville, MA, H. T. Kung *et al.* (eds) Computer Science Press, Rockville, Md (1981), 367–78

[43] NCR commercial note 45CG72, Geometric arithmetic parallel processor (1984)

[44] J. von Neumann, *Theory of Self-reproducing Automata*, University of Illinois Press, Urbana (1966)

[45] P. Quinton, Synthèse automatique d'architectures systoliques, Rapport IRISA (April 1985)

[46] Y. Robert, Block LU decomposition of a band matrix on a systolic array, *Int. J. Comput. Math.*, 17 (1985), 295–315

[47] Y. Robert and M. Tchuente, Réseaux systoliques pour des problèmes de mots, *RAIRO Informatique Théorique*, 19, 2 (1985), 107–23

[48] Y. Robert and M. Tchuente, Résolution systolique de systèmes linéaires denses, *RAIRO Modélisation et Analyse Numérique*, 19, 2 (1985), 315–26

[49] Y. Robert and M. Tchuente, An efficient systolic array for the 1D recursive convolution problem, *J. VLSI & Comput. Syst.*, 1, 4 (1986), 398–408

[50] Y. Robert and D. Trystram, Un réseau systolique orthogonal pour le problème du chemin algébrique, *C. R. Acad. Sci.*, 302, 1, 6 (1986), 241–4

[51] G. Rote, A systolic array algorithm for the algebraic path problem (shortest paths; matrix inversion), *Computing*, 34 (1985), 191–219

[52] R. Schreiber, Systolic arrays: high performance parallel machines for matrix computation, *Proc. Elliptic problem solvers II*, Academic Press, New York (1984), 187–94

[53] R. Schreiber and P. Kuekes, Systolic linear algebra machines in digital signal processing, in [34], 389–405

[54] E. Tiden, B. Lisper and R. Schreiber, Systolic arrays, Technical Report TRITA-NA-8315, The Royal Institute of Technology, Stockholm, Sweden (1983)

[55] J. D. Ullman, Computational aspects of VLSI, Chapter 5: Systolic algorithms, Computer Science Press, Rockville, Md (1984)

[56] T. Willey, R. Chapman, H. Yoho, T. S. Durrani and D. Preis, Systolic implementations for convolution, DFT and FFT, *Proc. IEEE*, 132, F, 6 (1985), 466–79

[57] H. C. Yung and C. R. Allen, A programmable VLSI array processor for IIR/FIR digital filtering, in *Digital Signal Processing*, V. Cappellini *et al.* (eds), Elsevier, North-Holland, Amsterdam (1984), 197–201

The systematic design of systolic arrays

9.1 Introduction

VLSI technology offers us exceptional opportunities to develop parallel computation for both special-purpose and general-purpose devices. Among the several approaches to parallel organisation that can take advantage of these new possibilities, the systolic array concept is particularly interesting. As characterised by Kung [5], a systolic array is a special-purpose parallel device, made out of a few simple cell types which are regularly and locally connected. Data circulate through these cells in a very regular fashion and interact whenever they meet, giving new results that are then sent to neighbouring cells. A number of systolic arrays have been designed for efficiently implementing algorithms related to various areas such as signal processing, image analysis and synthesis, numerical analysis and database management.

Systolic arrays are especially useful when one considers the evolution of integrated circuit technology, for the following reasons. First, the use of parallelism and pipelining allows much higher performance to be attained than can be achieved through sequential implementations. The price paid is the silicon area necessary to implement the processors. But it is a well known fact that the density of integrated devices is increasing an order of magnitude faster than the intrinsic speed of the technology. As a consequence, parallel and pipelined organisations can be expected to become more and more useful in the future as technology scales down. A second characteristic of systolic arrays is the geometric and timing regularity of their structure. The structure of systolic arrays is usually either a linear network or a two-dimensional grid or hexagonal network. Local connections help to solve one of the most difficult problems VLSI technology has to deal with, namely electrical delays in long wires. Moreover, the processing elements are supposed to be identical, or at least, based on a few models. As a result, the design time of, implementing such a device can be dramatically reduced by having a high replication factor. Finally the timing regularity of systolic arrays is of great importance relative to the overall control. One can think of systolic arrays as synchronous parallel devices, and moreover, the control of the operation can be made only by local decisions.

As the systolic array concept is expected to provide an efficient framework for the design of special-purpose VLSI architectures, it seems important to provide designers with methods that help them to explore various implementations of the same algorithm. Several alternatives can then be compared according to various criteria such as the size of the array, the complexity of the elementary cells, the throughput and the pipelining delay.

Attempts to synthesise systolic arrays can be classified into three categories. The first approach, called *functional transformation* consists in applying formal transformations to the mathematical expression of the algorithm [1,14]. These transformations introduce timing considerations through the use of the delay operator z^{-1}. In [14] it is shown how this approach can produce various systolic designs for numerical analysis algorithms. However, there is no systematic way to choose between all the possible transformations, and the method cannot be automated. A second approach, called *retiming*, is due to Leiserson and Saxe [7]. Starting from a design that is not systolic, that is where the data can have to go through an arbitrary amount of combinational logic at each cycle, the retiming method consists of applying a graph transformation which makes it possible to obtain an equivalent systolic design. Since the retiming method can be applied to an arbitrary design, it fails to capture the geometric regularity of systolic arrays. Moreover, the retiming method is not properly speaking a synthesis method, as the starting point is a particular implementation of the algorithm whose basic properties are not altered by the retiming. Finally, a third approach, called *dependence mapping*, aims at extracting the dependences between the variables of a program and at mapping the program onto a systolic array in such a way that the dependences are preserved. Attempts in this direction have been recently proposed by other workers: Moldovan [9] and Miranker and Winkler [8]. However, neither of these methods gives a systematic way to find the transformation. Also, since the algorithm has to be stated as a program, these methods cannot be applied to algorithms that deal with infinite flow of data, as commonly encountered in signal processing.

The method proposed here falls under the dependence mapping category. It is based on the possibility of expressing an algorithm as a set of uniform recurrence equations [4] over a convex polyhedral domain of $\mathbb{Z}^n$. When this is the case, it is possible under well-defined conditions to order the computations in such a way that the domain can be mapped onto a systolic array. In order to capture the essential properties of systolic systems, we require that the schedule and the mapping functions be essentially linear. This property has also the consequence that conditions under which these functions exist give rise to a systematic approach. Whenever a problem can be expressed as a system of uniform recurrence equations, a variety of systolic designs can be derived completely automatically.

The present paper is organised into six sections. In Section 9.2, we consider the principle of the method starting from the standard example of the convolution algorithm for which many systolic implementations exist. In Section 9.3, the problem is stated formally, and conditions are given under which a set of uniform recurrent equations can be scheduled and mapped onto a systolic array. We also gain some insight into the problem of finding optimal designs relative to criteria such as pipelining delay and throughput. Section 9.4 describes applications of the method to:

the convolution algorithm for which new designs called block-convolver and ring-convolver are derived,

the matrix product for which we show how two-dimensional systolic implementations can be found very simply using our approach.

Section 9.5 describes a generalisation of the method that gives it the flexibility to handle two-level pipelining. Section 9.6 summarises the method and gives some perspectives of this work.

9.2 Informal explanation of the method: the convolution product

Given a sequence $x(0)$, $x(1)$, $\ldots$, $x(i)$, $\ldots$ and coefficients $w(0)$, $w(1)$, $\ldots$, $w(K)$, the convolution algorithm consists of computing the sequence $y(0)$, $y(1)$, $\ldots$, $y(i)$, $\ldots$ where $y(i)$ is given by the equation

$$y(i) = \sum_{k=0}^{K} w(k)\, x(i - k) \qquad (9.1)$$

Equation (9.1) can be rewritten as

$$\forall i,\, 0 \leq i;\, \forall k,\, 0 \leq k \leq K: Y(i, k) = Y(i, k - 1) + w(k)\, x(i - k)$$
$$\forall i: 0 \leq i,\, Y(i, -1) = 0 \qquad (9.2)$$

where $Y(i, k)$ are the partial accumulated values for $y(i)$. We are interested here in finding systolic arrays that compute (9.2). For any integer coordinate point (i, k) lying in the domain $D = \{0 \leq i;\, 0 \leq k \leq K\}$ we have to compute a function $y_{out} = y_{in} + w_{in} x_{in}$, which will deliver $Y(i, k)$ provided y_{in}, w_{in}, and x_{in} are given correct values $Y(i, k - 1)$, $x(i - k)$ and $w(k)$. It should be observed that in equation (9.2) each computation makes use of two data items $w(k)$ and $x(i - k)$ that are common to some computations. In order to avoid reading the same data several times, it is possible to make $w(k)$ and $x(i - k)$ circulate from computation node to computation node. For example, $w(k)$, which is necessary for computing $Y(i, k)$, can be obtained as a by-product of the computation of $Y(i - 1, k)$. In the same manner, one can suppose that $x(i - k)$, which is needed for computing $Y(i, k)$, is given as a by-product of computing $Y(i - 1, k - 1)$. Note that other similar schemes are possible, and the first step of our method is to define precisely where values necessary for each computation are to be taken. Let us assume for example that input value $w(k)$ of node (i, k) is provided by node $(i - 1, k)$, and $x(i - k)$ by node $(i - 1, k - 1)$. Such a scheme can be formally expressed as the following system of equations:

$$\forall i: 0 \quad \leq i; \forall k: 0 \leq k \leq K$$

$$Y(i, k) = Y(i, k - 1) + W(i - 1, k) X(i - 1, k - 1)$$

$$W(i, k) = W(i - 1, k)$$

$$X(i, k) = X(i - 1, k - 1) \tag{9.3}$$

with the following initial conditions:

$$\forall i, 0 \leq i; \forall k, 0 \leq k \leq K:$$

$$Y(i, -1) = 0; \; W(-1, k) = w(k);$$
$$X(i - 1, -1) = x(i); \; X(-1, k - 1) = 0$$

Such a system of recurrent equations is said to be *uniform*, since computation at point (i, k) depends only on values computed at points that are obtained by a translation that does not depend on i or k (see [4]). Such a system can be represented by a graph such as that of Fig. 9.1. The nodes of this graph

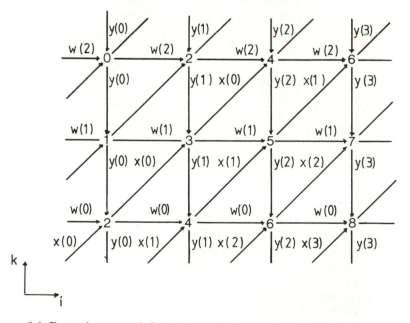

Figure 9.1 Dependence graph for the convolution product ($K = 2$)

represent the computations to be achieved and the edges represent values that are to be transmitted from one node to another. Such a graph can be more concisely abstracted as a domain D of $\mathbb{Z}^2$ (where $\mathbb{Z}$ denotes the set of integers) and a set of dependence vectors $\ominus = \{\theta_w, \theta_y, \theta_x\}$ that defines where node (i, k) has to take its input values, with $\theta_w = (1, 0)$, $\theta_y = (0, 1)$ and $\theta_x = (1, 1)$.

The next step of the method is to define a schedule for the computation by means of a timing function. A timing function is a mapping t from D to the set $\mathbb{Z}$ of integers such that if computation at point x of D depends on those at point y, then $t(x) > t(y)$. Such a function does not always exist. In the case of the above example, a very simple linear timing function exists (see Fig. 9.2), given by:

$$t(i, k) = i + k \tag{9.4}$$

According to this timing function, it is possible to solve (9.3) by successively executing computations for points $x \in D$ whose timing value $t(x)$ is $0, 1, \ldots, n, \ldots$ etc. The last step of our method is to define an architecture that supports these calculations. This is obtained by defining an allocation function a which maps D onto a finite set. Here also, a convenient way to derive simple solutions is to define the allocation function as a mapping from D to a finite domain of $\mathbb{Z}$ by means of a linear function. For our example, one can choose

$$a(i, k) = k \tag{9.5}$$

A very simple condition that such a function must satisfy is that the lines it defines are not parallel to the timing lines; when this condition is fulfilled, one

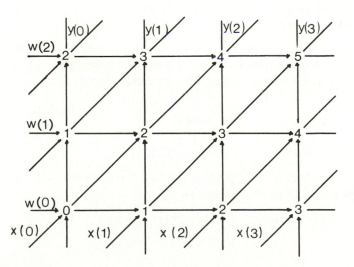

Figure 9.2 Timing function for the convolution product ($K = 2$)

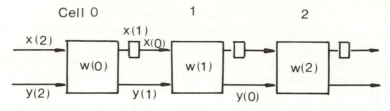

Figure 9.3 Systolic array for the convolution product

is sure that any processor has at most one computation to execute at a given instant. The functions t and a respectively given by (9.4) and (9.5) completely define a well-known systolic array depicted by Fig. 9.3. The whole domain D is mapped into 3 processors $p(0)$, $p(1)$ and $p(2)$. Each processor has to perform the computation attached to the points of D that are mapped onto it. The internal structure of each processor must be such that it supports the calculation described by equation (9.3). Communication among the processors is also completely defined, since processor k has to receive y, x and w from processors $k - a(\theta_y)$, $k - a(\theta_x)$, and $k - a(\theta_w)$ respectively. Since $a(\theta_y) = a(\theta_x) = 1$, we see that y and x come from processor $k - 1$, and since $a(\theta_w) = 0$, w_k stays in processor k. The timing of the operation of the systolic array is also well defined, and can be deduced from t and the value of the dependence vectors. Since $t(\theta_y) = 1$ and $t(\theta_x) = 2$, it is clear that the x values are going twice as slow as the y values.

In summary, the method described consists of three steps:

(i) Find a uniform recurrence equation system that solves the given problem
(ii) Find a timing function for this system.
(iii) Find an allocation function.

In the following, we will focus on steps (ii) and (iii). Although very important, step (i) will not be treated here. The interested reader can find in [2, 3, 12] some material covering this particular topic.

9.3   Formal description of the method

9.3.1   *Uniform recurrence equations (URE)*

Consider a subset D of $\mathbb{Z}^n$. Each point z of D is associated with a system of equations $E(z)$ given by

$$U_1(z) = f(U_1(z - \theta_1), \ldots, U_p(z - \theta_p))$$
$$U_2(z) = U_2(z - \theta_2)$$
$$\ldots$$
$$U_p(z) = U_p(z - \theta_p) \tag{9.6}$$

where f is a given p-variable function and θ_1, θ_2, ..., θ_p are vectors of $\mathbb{Z}^n$ called *dependence vectors*. Values $U_i(z)$ are assumed to be known outside of D, as defined by Karp *et al.* [4]. D is the *domain* of the URE, and the pair $(D, \ominus)$ where $\ominus = \{\theta_1, \theta_2, ..., \theta_p\}$ is called the *dependence graph of the URE*. A point $x \in D$ is said to *depend* on $y \in D$ if there exist $\theta_i \in \ominus$ such that:

$$x = y + \theta_i \tag{9.7}$$

We shall suppose that D is a set of integer coordinate points belonging to a *convex polyhedral domain* (CPD) of $\mathbb{R}^n$, that is a subset of $\mathbb{R}^n$ whose elements satisfy a set of linear inequalities with integer coefficients. D may thus be defined by a set of inequalities such as:

$$D = \{x \in \mathbb{Z}^n \mid Ax \le d\} \tag{9.8}$$

where A is a $q \times n$ matrix and d is a $q \times 1$ vector over $\mathbb{Z}$.

A few definitions and basic properties of convex sets are in order for the clarity of the following development (see [13]). Let $x_1, ..., x_m$ be points of $\mathbb{R}^m$. We say that $\sum_{i=1}^{m} \mu_i x_i$ is a *positive combination* of $x_1, ..., x_m$ if μ_i are nonnegative real numbers. $\sum_{i=1}^{m} \alpha_i x_i$ is a *convex combination* of $x_1, ..., x_m$ if it is a positive combination of $x_1, ..., x_m$ and $\sum_{i=1}^{m} \alpha_i = 1$.

Let C be a CPD. A point s of C is a *vertex* of C if s cannot be expressed as a convex combination of two different points of C. A vector r is said to be a *ray* of C if

$$\forall x \in C, \forall \mu \in \mathbb{R}^+, x + \mu r \in C \tag{9.9}$$

In other words, rays are the infinite directions of the CPD. A ray r of C is said to be an *extremal ray* if r cannot be expressed as a positive combination of other rays of C. A vector l of $\mathbb{R}^n$ is a *line* of C if

$$\forall x \in C, \forall \mu \in \mathbb{R}, x + \mu l \in C$$

If C has a line, it is said to be a *cylinder*. If C is not a cylinder, the set S of vertices of C is unique, and the set O of extremal rays of C is unique up to nonzero scalar multiples. C can then be defined as the subset of points x of $\mathbb{R}^n$ such that $x = y + z$, where y is a convex combination of vertices of S and z is a positive combination of rays of O. The pair (S, O) is called a *system of generators* of C.

A *face F* of C is a subset of C whose points meet at least one equality in equation (9.8). The vertices of C are particular faces of C. The faces of a CPD are themselves CPDs.

A CPD C is said to be n-dimensional if the smallest affine space generated by C is n-dimensional.

In the following, we shall assume that D is the subset of integer coordinate points of a CPD of $\mathbb{R}^n$ which is n-dimensional, and has *at most* one ray. Being unique, this ray is necessarily extremal.

As we have seen in the example of the convolution product, what we want to find is not a systolic array for one single domain, but more generally, a solution that can be extended to some family of domains. As an example, it is clear that the systolic array of Section 9.2 is extendible very naturally to any value of K. We thus introduce the following notion of problem.

Definition 1

Let I be a subset of the set $\mathbb{N}$ of natural numbers. A *problem* is a URE, and a family $F = (D_i)_{\{i \in I\}}$ of domains which have the same set of extremal rays O.

As an example, the URE given in Section 9.2 for the convolution is a problem when one considers the family of domains $F = (D_k)_{k \in \mathbb{N}}$ where:

$$D_k = \{(i, k) \mid i \geq 0 \text{ and } 0 \leq k \leq K\}$$

Definition 2

A family of domains F is $\ominus$-*extendible* if, for every $\theta \in \ominus$ and any positive integer $q \in \mathbb{N}$, there exists a domain D_i of the family that contains the segment $[x, x + q\theta]$ for some point x of D_i.

As an example, the first orthant $\{x \geq 0 \mid x \in \mathbb{Z}^n\}$ is extendible for any finite set $\ominus$ of vectors of $\mathbb{Z}^n$. In the same way, the family $F = (D_k)_{k \in \mathbb{N}}$ where $\ominus$ is the set of dependence vectors for the convolution and $D_k = \{(i, k) \mid 0 \leq i$ and $0 \leq k \leq K\}$ is extendible, although any finite subfamily of F with dependence vectors $\theta_x = (1, 1)$ and $\theta_y = (1, 0)$, does not have this property.

9.3.2   *Timing functions*

Throughout Sections 9.3 and 9.4, we assume that $\forall z \in D, U_1(z), \ldots, U_p(z)$ are evaluated simultaneously and their evaluation takes one unit of time. This hypothesis will be relaxed later on in Section 9.5. Consider a function t such that $t(z)$ is a time at which $E(z)$ can be computed consistently. Such a function will be called a *timing function*. Clearly if such a function exists, then at time $t(z)$ all the input arguments $U_i(z - \theta_i)$ must have been evaluated. On the other hand it is also desirable that the function be nonnegative, and that the number of computations to be performed simultaneously be bounded, so that the URE can be supported on a finite machine. We thus define a timing function in the following way.

Definition 3

A timing function for a dependence graph $(D, \ominus)$ is a mapping t from $\mathbb{Z}^n$ to $\mathbb{Z}$ such that:

  (i) t is nonnegative over D
 (ii) $\forall x, y \in D, t(x) > t(y)$ if x depends on y
(iii) there exists a positive integer M such that $\forall n \in \mathbb{N}, |\{x \mid x \in D$ and $t(x) = n\}| \leq M$ where $|E|$ denotes the cardinality of the set E.   (9.10)

In the following, we restrict ourselves to a particular class of timing functions called *quasi-affine timing functions* (QATF) which can be written

$$t(x) = \lfloor \lambda^T x - \alpha \rfloor, \ \lambda \in \mathbb{Q}^n \text{ and } \alpha \in \mathbb{Q} \tag{9.11}$$

where $\mathbb{Q}$ denotes the set of rational numbers and $\lfloor u \rfloor$ is the greatest integer less than or equal to u. We shall denote a QATF as a pair (λ, α).

Given a problem P with family domain $F = (D_i)_{\{i \in I\}}$, we denote by $T = (t_i)_{\{i \in I\}}$ the corresponding family of timing functions $t_i = (\lambda, \alpha_i)$.

The following theorem gives a necessary and sufficient condition that must be satisfied by a family of QATF:

Theorem 1
Let P be a problem with dependence set $\ominus$ and domain family $F = (D_i)_{\{i \in I\}}$. Let S_i be the set of vertices of D_i. If F is $\ominus$-extendible, then $T = (t_i)_{\{i \in I\}}$ is a family of QATF for P if and only if:

(i) $\forall \theta \in \ominus: \lambda^T \theta \geq 1$
(ii) $\forall r \in O: \lambda^T r > 0$
(iii) $\forall s \in S_i: \alpha_i \leq \lambda^T s$ (9.12)

The proof of Theorem 1 is given in Appendix 9.1.

The consequences of Theorem 1 are twofold. First, for any problem whose family of domains, is $\ominus$-extendible, we have a complete characterisation of the quasi-affine timing functions. Secondly, it can be seen that the set of conditions given by (9.12) gives a means to calculate such timing functions. Conditions (i) and (ii) show that λ belongs also to a CPD of $\mathbb{R}^n$. Let Λ be this CPD. In order to know if a problem has a timing function, it is sufficient to compute the system of generators of Λ. Any λ which is the sum of a convex combination of vertices of Λ and of a positive combination of rays of Λ will do. Then given a domain D_i, one can compute the scalar α_i using (iii).

Example
Consider the family of domains $D_K = \{(i, k) \mid 0 \leq i, 1 \leq k \leq K\}, K \geq 0$, and $\ominus = \{\theta_1, \theta_2, \theta_3\}$ where $\theta_1 = (4, 1)$, $\theta_2 = (1, 4)$ and $\theta_3 = (2, 2)$. Applying Theorem 1 gives:

$$4\lambda_1 + \lambda_2 \geq 1$$

$$\lambda_1 + 4\lambda_2 \geq 1$$

$$2\lambda_1 + 2\lambda_2 \geq 1$$

$$\lambda_1 > 0$$

The domain Λ is depicted in Fig. 9.4. Note that Λ has three vertices

$\sigma_1 = (1/3, 1/6)$, $\sigma_2 = (1/6, 1/3)$ and $\sigma_3 = (0, 1)$. If we choose $\lambda = \sigma_2$, the family of timing functions that results is given by

$$t_K(i, k) = \left\lfloor \frac{i}{6} + \frac{k}{3} \right\rfloor.$$

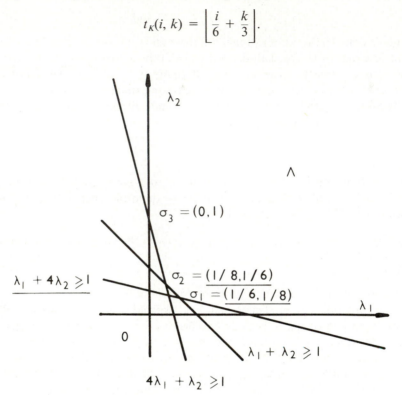

Figure 9.4 The convex polyhedral domain Λ defined by Theorem 1

9.3.3   *Optimal timing functions*

Theorem 1 gives the set of all possible QATF for a problem whose domain family is $\ominus$-extendible. Obviously, some timing functions are faster than others, in the sense that they optimise the throughput or the latency of the architecture: for example, it can be readily seen that if $t = (\lambda, \alpha)$ is a timing function for a given URE, then for any scalar $\mu > 1$, t is faster than $t' = (\mu\lambda, \alpha)$ which is also a QATF. In the following section, we define a partial order relation over QATFs which can be used for comparing them according to criteria such as pipelining delay, or throughput. We then give a theorem that allows timing functions that are optimal relative to these criteria to be found automatically.

Two criteria are usually used to compare different architectures supporting the same algorithm: the *throughput*, and the *pipelining delay*. The throughput is the average number of calculations performed per unit of time.

This parameter is clearly related to the 'speed' of the timing function along the ray r of the domain D when there exists one: the smaller $\lambda^T r$ is, the larger is the throughput of the machine. On the other hand, the pipelining delay is related to the speed of the timing function along some dependence vectors θ. In the convolution example the pipelining delay is proportional to $\lambda^T \theta_y$, which is the speed of t along the dependences between the successive partial values $Y(i, k)$. Note that the pipelining delay can involve several dependences, if the outputs depend simultaneously on several flows of input data.

Consider a problem over a family of domain $F = (D_i)_{\{i \in I\}}$ with dependence set $\ominus$. Let O be the set of common rays of F. Let $T = (\lambda, \alpha_i)$ and $T' = (\mu, \beta_i)$ be two families of timing functions for this problem.

Definition 4
We say that T is θ-*delay-faster* than T', and denote $T\theta - df\ T'$, if $\lambda^T\theta < \mu^T\theta$. If O is not void, we say that T is *throughput-faster* than T', and denote $T\ tf\ T'$ if $\lambda^T r < \mu^T r$.

According to Definition 4, finding the optimal timing functions relative to throughput or to the pipeline delay consists in finding the points $\lambda \in \Lambda$ which are minimal relative to the order relation $\theta - df$ or tf. The following theorem gives a characterisation of these points.

Theorem 2
Let (Σ, P) be a system of generators for Λ. Then T is minimal relative to $\theta - df$ (respectively tf) if λ belongs to a face of Λ generated by vertices of Σ which are minimal relatively to $\theta - df$ (respectively to tf), and rays ρ such that $\theta^T\rho = 0$ (respectively such that $r^T\rho = 0$).

Proof The proof results immediately from a well-known property of convex polyhedral domains. Given a convex polyhedral domain C, let $cl(C)$ be the closure of C, that is the CPD obtained when all the inequalities defining C are replaced by weak inequalities. Then if a linear functional has a lower bound on a convex polyhedral domain C, then it has a minimum value on $cl(C)$, and attains this minimum on a face of $cl(C)$. □

Theorem 2 gives a very simple way to determine where the optimal points are in Λ according to the above definitions. Minimal vertices of Λ can be found automatically, since a vertex σ is minimal if and only if it is minimal in the set Σ.

Note that if Λ has a unique vertex and contains this vertex, this vertex gives a solution that optimises all the criteria simultaneously and thus is delay-faster and throughput-faster than all the others. As we shall see in Section 9.5, this situation occurs very often in practice. It may also happen that Λ has no minimum point when Λ is not closed. This is the case when Λ has a unique vertex σ which meets $\sigma^T r = 0$ where r is the unique ray of D.

Since the throughput is proportional to $\frac{1}{\lambda^T r}$, this means that one can find timing functions with arbitrarily large throughput.

Example
Consider the example given at the end of Section 9.3.2 (Fig. 9.4). Choosing σ_1 would allow the pipelining delay along θ_2 and θ_3 to be minimised. Similarly, choosing σ_2 would minimise the pipelining delay along θ_1 and θ_2. Finally, it can be seen that any throughput can be achieved by taking λ as close as one wants to σ_3. The choice of λ clearly depends on the purpose of the systolic array.

9.3.4   *Allocation functions*

Once a QATF $t = (\lambda, \alpha)$ is found, it remains for us to map the computations onto a finite set of processors in such a way that each processor has at most one calculation to perform at a given time. Such a mapping will be called an *allocation function*. More precisely:

Definition 5
An allocation function for a computation graph $(D, \ominus)$ and a timing function $t = (\lambda, \alpha)$ is a mapping a from $\mathbb{Z}^n$ to some set E such that $a(D)$ is finite and $\forall x, y \in D$, $a(x) = a(y) \Rightarrow t(x) \neq t(y)$.

In order to obtain designs that have a good geometric regularity, we shall consider a particular class of allocation functions that we call *quasi-linear allocation functions* (QLAF).

Let us denote by $\mathbb{Z}_p$ the set of integers modulo p. By convention, $x \bmod 0 = x$ and $\mathbb{Z}_p = \mathbb{Z}$ if $p = 0$. Given vectors $P = (p_1, \ldots, p_n)$ and $X = (x_1, \ldots, x_n)$ of $\mathbb{Z}^n$, we write:

$$X \bmod P = (x_1 \bmod p_1, \ldots, x_n \bmod p_n)$$

Then function f from $\mathbb{Z}^n$ to $\mathbb{Z}^n$ is said to be quasi-linear (QLAF in the following), if there exists a vector P and a linear function g from $\mathbb{Z}^n$ to $\mathbb{Z}^n$ such that $f(x) = g(x) \bmod P$. A simple way to find a QLAF is to project the domain D along a conveniently chosen direction, as described in the following.

Considera nonnull vector u of $\mathbb{Z}^n$. Assume that $u_j \neq 0$. Suppose moreover that u is *primitive*, that is its components are relative prime number. Finally, suppose that $\lambda^T u > 0$, that is λ and u are not orthogonal. Let us denote by $e_i = (e_i^1, \ldots, e_i^n)$ the ith unit vector of $\mathbb{R}^n$, where e_j^i is δ_{ij} the Kronecker symbol. Clearly, the set $\left\{ \dfrac{e_1}{u_j}, \ldots, \dfrac{e_{j-1}}{u_j}, \dfrac{u}{u_j}, \dfrac{e_{j+1}}{u_j}, \ldots, \dfrac{e_n}{u_j} \right\}$ is a basis of $\mathbb{R}^n$. Let $x = (x_1, x_2, \ldots, x_n)$ be a point of $\mathbb{R}^n$ and let $y = (y_1, \ldots, y_n)$ be the representation of x over this new basis. A simple calculation shows that:

$$y_m = u_j x_m - u_m x_j \quad \text{if } m \neq j$$
$$y_j = x_j \tag{9.13}$$

Let $P = (p_1, \ldots, p_n)$ where:

$$p_m = 0 \qquad \text{if } m \neq j$$

$$p_j = \left\lceil \frac{1}{\lambda^T u} \right\rceil \tag{9.14}$$

where $\lceil x \rceil$ is the smallest integer greater than or equal to x. Let a be the mapping defined by:

$$a(x) = y \bmod P \tag{9.15}$$

Note that a is quasi-linear, since the mapping that associates y to x is linear. We then have the following theorem, whose proof is given in Appendix 9.2.

Theorem 3
If D has no ray, or if the ray r of D is parallel to u, then a is a QLAF for $(D, \ominus)$ and $t = (\lambda, \alpha)$.

Notice that if t has integer coefficients, then necessarily $\lambda^T u \geq 1$ and the mapping a reduces to a linear mapping of $\mathbb{Z}^n$ to $\mathbb{Z}^{n-1}$ that actually is a projection of $\mathbb{Z}^n$ along the direction u.

The above theorem gives a very simple method for deriving various designs from a unique URE, when D has no ray. When D has one ray r, the vector u must be parallel to r. However, Theorem 3 can be extended to handle the case when u is not parallel to r.

Let us denote by $M(u)$ the *integer width* of D along u, that is the greatest integer such that D contains a segment $[x, x + M(u)u]$. Let r be the unique ray of O. Assume that r is primitive and that r and u are not parallel. There exist integers j and k such that $r_k \neq 0$ and $u_j \neq 0$ and $r_k u_j - r_j u_k \neq 0$. Assume $j < k$ for example. Let

$$\Delta = r_k u_j - r_j u_k.$$

Clearly, the set

$$\left\{ \frac{e_1}{\Delta}, \ldots, \frac{u_j}{\Delta}, \frac{e_{j+1}}{\Delta}, \ldots, \frac{u_k}{\Delta}, \frac{e_{k+1}}{\Delta}, \ldots, \frac{e_n}{\Delta} \right\}$$

is a basis of $\mathbb{R}^n$. Given $x = (x_1, x_2, \ldots, x_n)$ in $\mathbb{R}^n$, let $y = (y_1, \ldots, y_n)$ be the representation of x over this new basis. A simple calculation shows that:

$$\forall m \neq j, k:$$

$$y_m = \Delta x_m - [r_m(x_k u_j - x_j u_k) + u_m(r_k x_j - r_j x_k)]$$

$$y_j = r_k x_j - r_j x_k$$

$$y_k = x_k u_j - x_j u_k$$

Let $P = (p_1, \ldots, p_n)$ where

$$p_m = 0 \text{ if } m \neq j, k$$

$$p_j = \left\lfloor \frac{1}{\lambda^T u} \right\rfloor$$

$$p_k = \left\lceil \frac{M(u)\lambda^T u + 1}{\lambda^T r} \right\rceil$$

Finally, let a be the mapping defined by $a(x) = y \bmod P$. We have the following theorem, whose proof is given in Appendix 3:

Theorem 4

The mapping a is a QLAF for $(D, \ominus)$ and $t = (\lambda, \alpha)$.

9.3.5 *Specification of the systolic array*

As we shall see in this section, a URE, a QATF $t = (\lambda, \alpha)$ and a QLAF a define completely a systolic array that supports the computation of the URE.

To each point π of $a(D)$ is attached one cell of the systolic array. To each point z of D, let us associate the program $P(z)$ defined by

begin for j := 1 **to** p **do**
   if $a(z + \theta_j) \in a(D)$ **then**
      read $U_j(z + \theta_j)$ from cell $a(z + \theta_j)$
   else read $U_j(z + \theta_j)$ from memory;

for j := 1 **to** p **do**
   compute $U_j(z)$ according to the URE;
for j := 1 **to** p **do**
   if $a(z - \theta_j) \in a(D)$ **then**
      send $U_j(z)$ to cell $a(z - \theta_j)$
      with a delay $\Delta_j(z) = t(z - \theta_j) - t(z)$
   else write $U_j(z)$ into memory
end

Cell π of the array has to perform program $P(z)$ at time $t(z)$, for all z lying in $a^{-1}(\pi)$. One can prove that:

Theorem 5

  (i) $\forall \pi, \forall n \geq 0$, there exists at most one z such that $t(z) = n$
  (ii) $\forall z \in D, \forall \theta_j \in \theta, U_j(z + \theta_j)$ is available at time $t(z)$
  (iii) Let $\pi = (\pi_1, \ldots, \pi_n) \in a(D)$. Then $\forall z \in a^{-1}(\pi), \forall \theta_i \in \ominus: a(z + \theta_i) = (\pi + a(\theta_i)) \bmod P$ and $a(z - \theta_i) = (\pi - a(\theta_i)) \bmod P$
  (iv) Suppose that D has one ray. Let $(P_n)_{\{n \in \mathbb{N}\}}$ be the sequence of programs performed by π. There exists k such that $(P_n)_{\{n > k\}}$ is periodic.

Proof :

Parts (i) and (ii) of the theorem result immediately from Theorem 1 and Theorem 3. Parts (iii) and (iv) can be established by examining the values taken by $a(z + \theta_j)$, $a(z - \theta_j)$ and $\Delta_j(z)$ for every $z \in a^{-1}(\pi)$. □

The consequences of Theorem 5 are the following:

Each cell has at most one program to perform at a given time (i) and input values of this program have previously been calculated (ii);

Each cell must be connected to only a finite number of other cells (iii). Moreover, condition (iii) can be interpreted as a property of locality and extensibility of the design, when one considers a problem rather than a unique URE (for example, the convolution product with coefficient vectors of any size).

Finally, a cell must contain only a finite number of different programs and can determine what program to perform using a finite state control mechanism (iv).

9.4   Applications

This section of the paper shows how the above described method can be applied to well-known problems. We successively investigate various systolic designs for convolution, and matrix product.

9.4.1   *Convolution*

There exist many solutions for this problem. In the present section, we show how classical systolic arrays and new ones can be derived.

In Section 9.2, we considered a first uniform recurrence system for the convolution equations. Let us now suppose that values w_k are kept from node $(i - 2, k)$ instead of from node $(i - 1, k)$ as in the previous form. We then obtain the new URE:

$$Y(i, k) = Y(i, k - 1) + W(i - 2, k)X(i - 1, k - 1)$$

$$X(i, k) = X(i - 1, k - 1)$$

$$W(i, k) = W(i - 2, k) \tag{9.16}$$

whose dependence graph $(D_K, \ominus)$ is given (see Fig. 9.5) by:

$$D_K = \{0 \leq k \leq K; 0 \leq i\}$$

and

$$\ominus = \{\theta_y = (0, 1), \theta_x = (1, 1), \theta_w = (2, 0)\}$$

Theorem 1 shows that a necessary and sufficient condition for (λ, α_k) to be a QATF is:

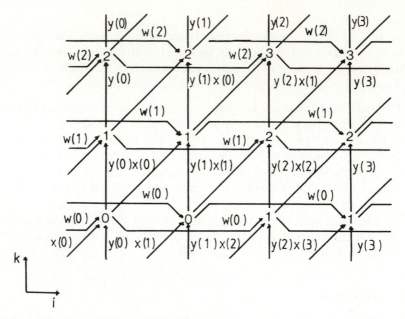

Figure 9.5 Another dependence graph for the convolution product. The timing function is $t(i, k) = \left[\dfrac{i}{2} + k \right]$.

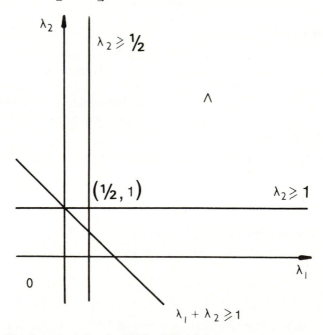

Figure 9.6 The domain Λ for the URE of Fig. 9.5

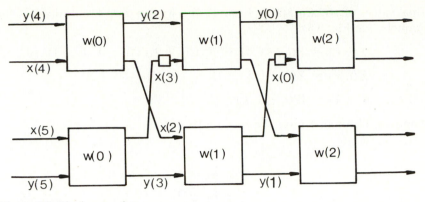

Figure 9.7 Block-convolver

$$\lambda_2 \geq 1; \lambda_1 + \lambda_2 \geq 1; 2\lambda_1 \geq 1$$

and

$$\alpha_K \leq \min \{\lambda^T s | s \in S_K\}$$

where $S_K = \{(0, 0), (0, K)\}$ is the set of vertices of D_K. Figure 9.6 depicts the corresponding domain Λ. The timing function t_K defined by

$$t_K(i, k) = \left\lfloor \frac{i}{2} + k \right\rfloor$$

meets these conditions and is throughput and delay optimal according to Theorem 2.

As shown by Fig. 9.5, this QATF allows us to compute $2(K + 1)$ values instead of $(K + 1)$ in the design that was given in Section 9.2. If we choose to project D along the i-axis, that is if we choose $u = (2, 0)$, we then obtain by applying Theorem 3:

$$a(i, k) = (i \bmod 2, k)$$

which results in the systolic array depicted by Fig. 9.7. We call this design a *block-convolver* since two values x_i are input simultaneously. Notice that this design can be extended without difficulty to the parallel computation of $n(K + 1)$ values, by using a w dependence $\theta_w = (n, 0)$.

Other solutions can be found by modifying the initial recurrent system of equations (9.2). For example, another way to serialise the computation of the y_i's is to proceed 'backwards', that is by summing up terms involving w_k by decreasing values of k. One has then

$$\forall k: 0 \leq k \leq K, \forall i: 0 \leq i$$

$$Y(i, k) = Y(i, k + 1) + w(k)x(i - k)$$

$$Y(i, K + 1) = 0 \tag{9.17}$$

Final values of $y(i)$ are $Y(i, 0)$. From (9.16), we can derive several URE, for example (see Fig. 9.8):

$$Y(i, k) = Y(i, k + 1) + W(i - 1, k)X(i - 1, k - 1)$$

$$X(i, k) = X(i - 1, k - 1)$$

$$W(i, k) = W(i - 1, k)$$

Applying Theorem 1 gives the constraints over λ:

$$-\lambda_2 \geq 1;$$

$$\lambda_1 \geq 1;$$

$$\lambda_1 + \lambda_2 \geq 1$$

This domain has a single vertex $\lambda = (2, 1)$. If we choose this value, α_K is given by 9.12(iii) and reaches its minimum value at the vertex $(0, K)$ so that

$$t_K(i, k) = 2i - k + K = 2i + (K - k)$$

is a possible timing function which is delay and throughput optimal. By projecting D along the i-axis, we obtain the QLAF $a(i, k) = (0, k)$ which gives the well-known design of Fig. 9.9.

Consider now the projection defined by $u = (1, -1)$. Applying Theorem 4 by replacing the canonical basis of $\mathbb{R}^n$ by $\{r, u\}$, gives:

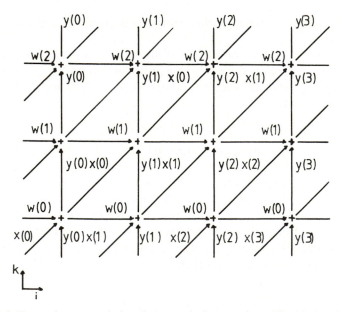

Figure 9.8 Dependence graph for the convolution product. The timing function is $t(i, k) = 2i + (K - k)$.

$$a(i, k) = (i + k \bmod 4, 0)$$

since

$$\left\lceil \frac{M(u)\lambda^T u + 1}{\lambda^T t} \right\rceil = 4 \text{ and } \left\lfloor \frac{1}{\lambda^T u} \right\rfloor = 1$$

Figure 9.10 shows the systolic structure that is obtained once the cells are arranged on a ring. This design, called a *ring-convolver* has 4 cells. Cell i reads values y from cell $i + 1 \bmod 4$ or from the host. It reads values x from cell $i - 2 \bmod 4$ or from the host. It reads w from cell $i - 1 \bmod 4$. It sends y

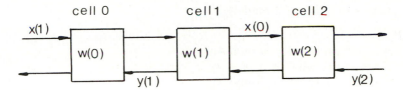

Figure 9.9 Systolic array for the convolution product

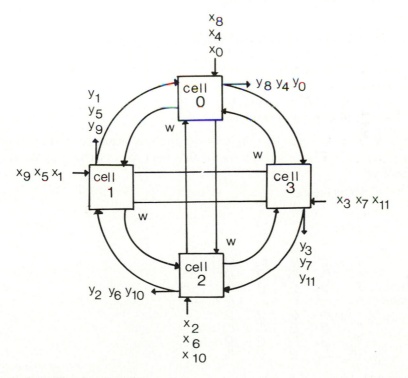

Figure 9.10 The ring-convolver

either to cell $i - 1 \bmod 4$ or to the host, and x either to cell $i + 2 \bmod 4$ or to the host. Finally, notice that the w-connections between cell i and cell $i + 1 \bmod 4$ have a delay of two units of time. Each cell indefinitely executes a sequence of eight cycles $P_1, Q, Q, P_2, Q, Q, P_3, Q$, where Q denotes the idle cycle and P_1, P_2, P_3 are defined by:

P_1:
  read y from the host;
  read w from cell $(i - 1 \bmod 4)$;
  read x from cell $(i - 2 \bmod 4)$;
  send $y + wx$ to cell $(i - 1 \bmod 4)$;
  send w to cell $(i + 1 \bmod 4)$;
  send x to cell $(i + 2 \bmod 4)$;

P_2:
  read y from cell $(i - 1 \bmod 4)$;
  read w from cell $(i + 1 \bmod 4)$;
  read x from cell $(i - 2 \bmod 4)$;
  send $y + wx$ to cell $(i - 1 \bmod 4)$;
  send w to cell $(i + 1 \bmod 4)$;
  send x to cell $(i + 2 \bmod 4)$;

P_3:
  read y from cell $(i + 1 \bmod 4)$;
  read w from cell $(i - 1 \bmod 4)$;
  read x from the host;
  send $y + wx$ to the host;
  send w to cell $(i + 1 \bmod 4)$;
  send x to cell $(i + 2 \bmod 4)$;

Depending on their number, the cells execute this sequence starting at one particular cycle. Figure 9.11 shows the first ten cycles executed by the ring-convolver.

		Cell number		
		0	1	2	3
	0	—	—	P_1	—
	1	—	P_2	Q	—
	2	P_3	Q	Q	P_1
	3	Q	Q	P_2	Q
	4	P_1	P_3	Q	Q
Cycle	5	Q	Q	Q	P_2
number	6	Q	P_1	P_3	Q
	7	P_2	Q	Q	Q
	8	Q	Q	P_1	P_3
	9	Q	P_2	Q	Q
	10	P_3	Q	Q	P_1

Figure 9.11 Activity of the cells of the ring–convolver during the first ten cycles

9.4.2 Matrix product

Consider two $N \times N$ matrices A and B and let $C = AB$. We have

$$\forall i, j: 1 \leq i \leq N, 1 \leq j \leq N$$

$$c_{ij} = \sum_{k=1}^{N} a_{ik} b_{kj} \tag{9.18}$$

As in the case of the convolution product, a natural way to serialise (9.18) is to put

$$\forall i, j, k: 1 \leq i \leq N, 1 \leq j \leq N, 1 \leq k \leq N$$

$$C(i, j, k) = C(i, j, k - 1) + a_{ik} b_{kj}$$

$$C(i, j, 0) = 0$$

It is clear that a_{ik} is used by nodes (i, j, k) where $1 \leq j \leq N$ and b_{kj} is used by nodes (i, j, k) where $1 \leq i \leq N$. A natural mode of circulation of these data is that node (i, j, k) takes a_{ik} from node $(i, j - 1, k)$ and b_{kj} from node $(i - 1, j, k)$. This leads to the following URE:

$$C(i, j, k) = C(i, j, k - 1) + A(i, j - 1, k)B(i - 1, j, k)$$

$$A(i, j, k) = A(i, j - 1, k)$$

$$B(i, j, k) = B(i - 1, j, k) \tag{9.19}$$

The dependence graph for (9.19) is $(D_N, \ominus)$ where

$$D_N = \{(i, j, k) | 1 \leq i \leq N, 1 \leq j \leq N, 1 \leq k \leq N\}$$

is a cube of $\mathbb{R}^n$ and

$$\ominus = \{\theta_c = (0, 0, 1), \theta_a = (0, 1, 0), \theta_b = (1, 0, 0)\}$$

Applying Theorem 1 gives the following constraints over $\lambda = (\lambda_1, \lambda_2, \lambda_3)$:

$$\lambda_1 \geq 1; \lambda_2 \geq 1; \lambda_3 \geq 1$$

The domain Λ defined by these inequalities has a single vertex $\lambda = (1, 1, 1)$, that can be chosen for the timing function. $t_N(i, j, k) = i + j + k - 3$ is a possible timing function.

There are numerous way to map the domain D using Theorem 3. We consider here only two designs, obtained using projection directions $u = (0, 0, 1)$, and $u = (1, 1, -1)$.

First Case: $u = (0, 0, 1)$

This first design consists in projecting the domain along the k-axis. The result is depicted in Fig. 9.12. The allocation function obtained using Theorem 3 is:

$$a(i, j, k) = (i, j, 0)$$

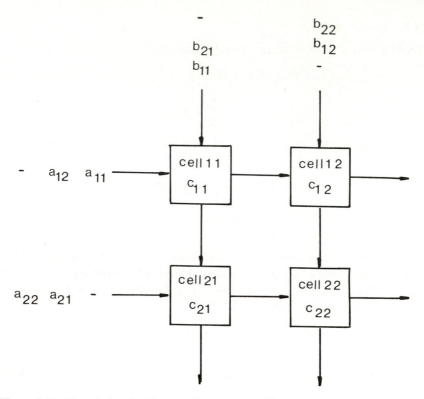

Figure 9.12 First design for the matrix product; coefficients c_{ij} stay in cell (i, j). The network contains N^2 cells.

and maps D_N onto the square $\{1 \leq i \leq N; 1 \leq j \leq N\}$. This design contains thus N^2 cells, and is the smallest one that can be obtained from D_N. However, it suffers from the main drawback that the results are not output naturally from the array, since they do not move during the computation.

Second case: $u = (1, 1, -1)$
Applying Theorem 3 gives the following allocation function (obtained by replacing the basis vector $(1, 0, 0)$ with u):

$$a(i, j, k) = (j - 1, i + k, 0)$$

The resulting systolic array is shown in Fig. 9.13. The domain $a(D_N)$ contains $3N^2 - 3N + 1$ cells. This design is similar to that described by Kung [6] but allows one value to be computed every cycle rather than every three cycles. (Notice, however, that Kung's design is more interesting when band matrices are considered, as the domain is truncated and the projection along vector $(1, 1, 1)$ has less processors.)

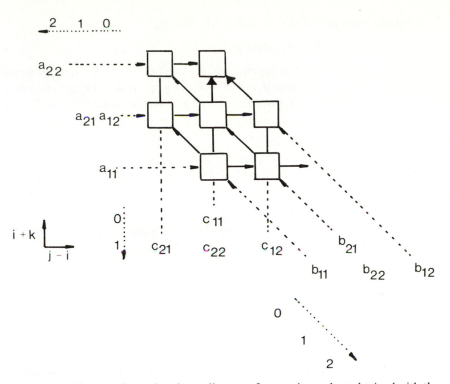

Figure 9.13 The two-dimensional systolic array for matrix product obtained with the projection vector $u = (1, 1, -1)$

9.5 Generalisation

In this section we generalise the notion of Uniform Recurrence Equations introduced in Section 9.3.1 in such a way that they can handle problems such as two-level pipelining. Our purpose is to provide a framework flexible enough to handle solution refinements.

A *Generalised Uniform Recurrent System* is a system of equations $E(z)$ given by:

$$\forall m, \ 1 \le m \le p: U_m(z) = f_m(U_{m_1}(z - \theta_{m_1}), \ldots, U_{m_k}(z - \theta_{m_k}))$$

where $m_1, m_2, \ldots, m_k$ belong to $[1, m]$ and θ_j are vectors of $\mathbb{Z}^n$. Points z are again supposed to be integer-coordinate points of a convex polyhedral domain D of $\mathbb{R}^n$ which is *n*-dimensional and has at most one ray.

We make the following assumption on the operational semantics of the system. We suppose that the evaluation of equation number m at point z starts when all the arguments of its right-hand side are available, and that the time τ_i to perform this evaluation is constant and is a positive integer. In other words, we suppose that the evaluation of function f_j is *atomic*.

Consider now the subset $\tilde{D}$ of $\mathbb{R}^{n+1}$ defined by:

$$\tilde{D} = \{(z, m)\,|\,z \in D \text{ and } 1 \leq m \leq p\}$$

A point $(x, m) \in \tilde{D}$ is said to depend on $(y, l) \in \tilde{D}$ if $U_l(z - \theta_l)$ appears as an argument of function f_m, and if $x = y + \theta_l$. A generalised timing function is an m-tuple $t = (t_1, \ldots, t_p)$ where each t_m is a mapping from $\mathbb{Z}^n$ to $\mathbb{Z}$ such that:

(1) t_m is non negative over D;
(2) if (x, m) depends on (y, l) then $t_m(x) \geq t_l(y) + \tau_l$
(3) there exists $M > 0$ such that $\forall n \in \mathbb{N}$, $|\{x\,|\,x \in D \text{ and } t(x) = n\}| \leq M$

(9.20)

A generalized timing function will be said to be quasi-affine if all the functions t_m are quasi-affine and have the same slope λ, that is there exist $\lambda \in \mathbb{Q}^n$, and rational numbers $\alpha_1, \ldots, \alpha_p$ such that:

$$\forall m, 1 \leq m \leq p\colon t_m(z) = \lfloor \lambda^T z - \alpha_m \rfloor$$

We shall denote a generalised QATF as $(\lambda, \alpha_1, \ldots, \alpha_p)$. The following theorem gives a set of conditions which are sufficient for t to be a generalised timing function.

Theorem 6
The function $t = (\lambda, \alpha_1, \ldots, \alpha_p)$ is a generalised QATF for a generalised URE if:

(1) $\forall m, 1 \leq m \leq p$, $\forall l$ such that U_l is an argument of f_m:

$$\lambda^T \theta_l \geq \alpha_m - \alpha_l + \tau_l$$

(2) if D has a ray r, then $\lambda^T r > 0$
(3) $\forall m, 1 \leq m \leq p, \forall s \in S, \alpha_m \leq \lambda^T s$ (9.21)

Proof
Conditions (9.21.2) and (9.21.3) imply that each function t_m is nonnegative over D. Property (9.20.3) results from (9.21.3). It thus remains to prove that $t_m(x) \geq t_l(y) + \tau_l$ whenever (x, m) depends on (y, l). But (x, m) depends on (y, l) if there exists an argument $U_l(x - \theta_l)$ of f_m such that $x = y + \theta_l$. Therefore:

$$t_m(x) = \lfloor \lambda^T x - \alpha_m \rfloor = \lfloor \alpha^T y + \alpha^T \theta_l - \alpha_m \rfloor$$

But, from (9.21.1) we have:

$$\lfloor \lambda^T y + \lambda^T \theta_l - \alpha_m \rfloor \geq \lfloor \lambda^T y - \alpha_l + \tau_l \rfloor = \lfloor \lambda^T y - \alpha_l \rfloor + \tau_l \qquad \square$$

Theorem 6 shows that the vector λ and the scalars α_m which lie in the convex polyhedral domain of $\mathbb{R}^{n+m}$ defined by (9.21) are solutions to the problem. As a consequence, we can determine possible values of these parameters by a fully constructive method. As far as allocation functions are concerned, it is obvious that Theorems 3 and 4 may be applied directly, since they involve only values λ. Finally the internal structure of each cell can be derived from the generalised timing function and the allocation function in a very similar way as explained in Section 9.3.4. The only difference lies in the fact that each processor may be thought of as partitioned into m sub-processors, one for each equation of the URE.

As an example, let us show how a two-level systolic array for the convolution can be produced using this generalised method. Let us start again from the URE:

$$\forall i, 0 \quad \leq i, \forall k, 0 \leq k \leq K:$$

$$Y(i, k) = Y(i, k - 1) + W(i - 1, k)X(i - 1, k - 1)$$

$$W(i, k) = W(i - 1, k)$$

$$X(i, k) = X(i - 1, k - 1)$$

Suppose now that we want to refine the internal structure of each cell of the array, in such a way that the adder and the multiplier are separated. In order to do so we just need to rewrite the computation of Y by introducing a separate term P for the product of W and X. We thus get:

$$\forall i, 0 \quad \leq i, \forall k, 0 \leq k \leq K:$$

$$Y(i, k) = Y(i, k - 1) + P(i, k)$$

$$P(i, k) = W(i - 1, k)X(i - 1, k - 1)$$

$$W(i, k) = W(i - 1, k)$$

$$X(i, k) = X(i - 1, k - 1) \tag{9.22}$$

Let us denote by α_Y, α_P, α_W and α_X the constant coefficients of the timing function, and by τ_Y, τ_P, τ_W and τ_X the time to compute the terms Y, P, W and X respectively. Using condition (9.21.1) of Theorem 5 we get:

$$\lambda_2 \geq \tau_y; \; 0 \geq \alpha_y - \alpha_p + \tau_p$$

$$\lambda_1 \geq \alpha_p - \alpha_w + \tau_w;$$

$$\lambda_1 + \lambda_2 \geq \alpha_p - \alpha_x + \tau_x$$

$$\lambda_1 \geq \lambda_w; \; \lambda_1 + \lambda_2 \geq \tau_x$$

and using conditions (9.21.2) and (9.21.3), we obtain:

$$\lambda_1 > 0$$

$$\alpha_y \leq 0; \; \alpha_y \leq \lambda_2 K$$

$$\alpha_P \leq 0;\ \alpha_P \leq \lambda_2 K$$

$$\alpha_W \leq 0;\ \alpha_W \leq \lambda_2 K$$

$$\alpha_X \leq 0;\ \alpha_X \leq \lambda_2 K$$

It turns out that the convex domain defined by the above inequalities has only two vertices which are:

$$\lambda_1 = 1;\ \lambda_2 = 1;\ \alpha_y = -1;\ \alpha_p = \alpha_x = \alpha_w = 0$$

and

$$\lambda_1 = 1;\ \lambda_2 = 1;\ \alpha_y = -1;$$

$$\alpha_X = -1;$$

$$\alpha_P = \alpha_W = 0$$

If we take any of these solutions, and project D along the i-axis, (i.e. using the allocation function $a(i, k) = 0, k$), we obtain the systolic array shown in Fig. 9.14.

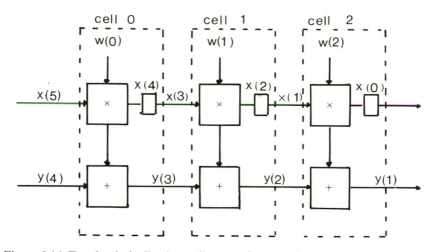

Figure 9.14 Two-level pipelined systolic array for convolution

9.6 Conclusion

This chapter has described a powerful method for the systematic design of systolic arrays. The method consists in deriving uniform recurrence equations for a given problem, the finding a timing function for the computations, and finally projecting the domain of computation along a direction which is not parallel to the hyperplanes defined by the timing function. Conditions upon which which a timing function and an allocation function exist are given. These conditions are constructive and allow a systolic array that implements a given algorithm to be built step by step, provided that some simple conditions are met. Nontrivial applications of this method have been described, showing that it can be used for finding entirely new designs.

The method described above is currently used as the basis of a computer aided design (CAD) system for systolic arrays [11]. This CAD system, named DIASTOL, is intended to provide a designer with interactive graphics facilities and various simulation tools, which are necessary for the fast and reliable design of special-purpose architectures. DIASTOL produces designs that are necessarily correct and consequently do not need to be proved, provided that the uniform recurrence system of equations has been proved to be formally equivalent to the mathematical equations of the problem.

Among the numerous problems that still need to be investigated, let us cite only two that seem fundamental. The first is to explore formally which class of algorithms may be expressed by means of URE, and to define formal transformations that allow equivalent URE's to be derived from the algorithm. Another problem to be solved is the evaluation of the solutions that are obtained by such a method. This is a very difficult problem, since a quality measure must include a number of parameters such as size, speed and memory bandwidth, which need to be evaluated not only formally, but also relative to the function that a systolic array has to perform, that is the system in which it is to be included.

9.7 References

[1] D. Cohen, Mathematical approach to iterative computation networks, in *Proc. 4th Symp. on Computer Arithmetic* (1978), 226–38

[2] J. Fortes and D. Moldovan, Data broadcasting in linearly scheduled array processors, in *Proc. 11th Annual Symp. on Computer Architecture, Ann Arbor, June 1984*

[3] P. Gachet, B. Joinnault and P. Quinton, Synthesizing systolic arrays using DIASTOL, *Int. workshop on Systolic Arrays, Oxford (UK), July 1986*

[4] R. M. Karp, R. E. Miller and S. Winograd, The organization of computations for uniform recurrence equations, *J. Assoc. Comput. Mach.*, 14, 3(1967), 563–90

[5] H. T. Kung, Why systolic architectures? *Computer*, 15, 1 (1982), 37–46

[6] H. T. Kung and C. E. Leiserson, Systolic arrays (for VLSI), in Chap. 8 of *Introduction to VLSI Systems*, C. Mead and L. Conway (eds), Addison-Wesley, Reading, Mass. (1980)

[7] C. E. Leiserson and J. B. Saxe, Optimizing synchronous circuits, in *Proc. 22nd IEEE Symp. on Foundations of Computer Science* (1981), 23–36

[8] W. L. Miranker and A. Winkler, Spacetime representations of systolic computational structures, IBM Research Report RC 9775 (1982)

[9] D. I. Moldovan, On the design of algorithms for VLSI systolic arrays, *Proc. IEEE*, 71, 1, (1983), 113–20

[10] P. Quinton, The systematic design of systolic arrays, IRISA Research Report No. 193 (April 1983)

[11] P. Quinton and P. Gachet, DIASTOL Users Manual: Preliminary Version, IRISA Research Report (1984)

[12] P. Quinton, B. Joinnault and P. Gachet, A new matrix multiplication systolic array, *Int. Symp. on Parallel Algorithms and Architectures, Luminy (France), April 1986*

[13] R. T. Rockafellar, *Convex Analysis*, Princeton University Press, Princeton, NJ (1970)

[14] U. Weiser and A. Davis, A wavefront notation tool for VLSI design, in *VLSI Systems and Computations*, H. T. Kung, B. Sproull and G. Steele (eds), Computer Science Press, Rockville, Md (1981), 226–34

Acknowledgments

I would like to acknowledge the help of many people during the above reported study. The proof of Theorem 1 is partly due to Pascal Quinton, who also contributed greatly to the formal explanation of many informal ideas. Thanks also to Philippe Darondeau, with whom I had a number of discussions about this work. Finally, thanks to Françoise André, Albert Benveniste, Arnold Rosenberg, and Laurent Kott who read this paper very carefully and corrected a number of errors.

Appendix 9.1

Proof of Theorem 1

Necessary condition   Note that (9.12.3) is obviously met, since otherwise $t_i(x)$ would be negative at least at one vertex of some domain D_i and would not be a timing function for this particular domain D_i.

Suppose now that (9.12.2) is not true. If $\lambda^T r < 0$, by definition of a ray (9.9):

$$\forall \mu > 0, \forall x \in D: x + \mu r \in D$$

It is then clear that there must exist a domain D_i, a point x in D_i and a positive real number μ, such that $\lambda^T(x + \mu r) - \alpha_i < 0$, which is not consistent with the definition of a timing function. On the other hand, if $\lambda^T r = 0$, the number of points computed simultaneously cannot be bounded, since all the points lying on a ray of D have the same timing value.

Let us now prove that (9.12.1) is true. Let $\theta \in \ominus$. Note first that, since F is $\ominus$-extendible, there exist D_i and $x \in D_i$ such that $x - \theta \in D_j$. Thus

$$t_i(x) - t_i(x - \theta) \geq 1$$

or

$$[\lambda^T x - \alpha_i] - [\lambda^T x - \alpha_i - \lambda^T \theta] \geq 1$$

As a consequence, $\lambda^T \theta > 0$.

To prove that $\lambda^T \theta \geq 1$, we shall prove that every rational number that is greater than or equal to $\lambda^T \theta$ is also greater than or equal to 1. This will suffice to prove that $\lambda^T \theta \geq 1$, since

$$\lambda^T \theta = \inf\{\alpha \in \mathbb{Q} | \alpha \geq \lambda^T \theta\}$$

Let p/q be a rational number such that $p/q \geq \lambda^T \theta$. Since F is $\ominus$-extendible, there exist D_i and $x \in D_i$ such that $[x, x - q\theta] \subset D_i$. We then have

$$t_i(x) \geq t_i(x - \theta) + 1 \geq \dots \geq t_i(x - q\theta) + q$$

but since $\lambda^T \theta \leq p/q$:

$$t_i(x - q\theta) = \lfloor \lambda^T x - \alpha_i - q\lambda^T \theta \rfloor \geq \lfloor \lambda^T x - \alpha_i - p \rfloor = t_i(x) - p$$

We thus have

$$t_i(x) - q \geq t_i(x - q\theta) \geq t_i(x) - p$$

which implies that $p \geq q$ and thus $p/q \geq 1$.

Sufficient condition

Let us first prove that t_i is positive over D_i. We assume that D_i has a ray r (the proof is similar if D has no ray). Every $x \in D_i$ can be written as the sum of a convex combination of the vertices $s \in S_i$ of D_i and of a positive combination of rays of D_i. Thus:

$$x = \sum_{j=1}^{u} \beta_j s_j + \mu r, \text{ where } s_j \in S_i, \beta_j \geq 0, \sum_{j=1}^{u} \beta_j = 1, \mu \geq 0$$

thus

$$\lambda^T x - \alpha_i = \sum_{j=1}^{u} \beta_j \lambda^T s_j + \mu \lambda^T r - \alpha_i$$

But from (9.12.3), we have

$$\sum_{j=1}^{u} \beta_j \lambda^T s_j \geq \sum_{j=1}^{u} \beta_j \alpha_i = \alpha_i$$

Thus

$$\lambda^T x - \alpha_i \geq \mu \lambda^T r$$

But since λ satisfies condition (9.12.2), we can conclude that

$$\forall x \in D_i, \lfloor \lambda^T x - \alpha_i \rfloor \geq 0$$

which implies that t_i is nonnegative over D_i.

Let us now prove that if x and y are in D_i and x depends on y, then $t_i(x) \geq t_i(y) + 1$. Since x depends on y, there exists $\theta_j \in \theta$ such that:

$$x = y + \theta_j$$

We can write:

$$t(x) = \lfloor \lambda^T x - \alpha_j \rfloor = \lfloor \lambda^T (y + \theta_j) - \alpha_i \rfloor$$
$$= \lfloor \lambda^T y - \alpha_i + \lambda^T \theta_j \rfloor \geq \lfloor t(y) + \lambda^T \theta_j \rfloor$$

But (9.12.1) implies that $\lambda^T \theta_j \geq 1$ and thus $t_i(x) \geq t_i(y) + 1$.

Finally we have to prove that the number of points of D_i scheduled at a given time n is bounded. Suppose this is not true for some domain D_i. There exists then an infinite number of points in D_i that have the same timing value n, and since t is quasi-linear, D_i has one ray ρ that must satisfy $\lambda^T \rho = 0$. Since O contains at most one ray r, clearly we must have $r = \mu \rho$ for some scalar μ. Thus, $\lambda^T r = 0$, which contradicts (9.12.2). $\square$

Appendix 9.2

Proof of Theorem 3

The proof of this theorem makes use the following result:

Lemma 1   Let x and y be two points of D. Then $a(x) = a(y)$ if and only if there exists an integer k such that $x = y + kp_j u$.

Proof   Suppose that $a(x) = a(y)$. Let $x = y + z$. From (9.15) and (9.13):

$$\forall m \neq j: u_j z_m - u_m z_j = 0 \tag{9.23}$$

and

$$z_j \bmod p_j = 0 \tag{9.24}$$

From (9.24), there exists an integer l such that

$$z_j = l p_j \tag{9.25}$$

Substituting (9.25) into (9.23), we get

$$\forall m \neq j: z_m = \frac{l u_m}{u_j p_j}$$

But since $z \in \mathbb{Z}^n$ and u is reduced, u_j must divide l. Let $k = \dfrac{l}{u_j}$. We thus have $z = kp_j u$ which establishes the result. Conversely, if $x = y + kup_j$, a simple calculation shows that $a(x) = a(y)$.   □

Proof of Theorem 3

We must prove that $a(D)$ is a finite set, and that $a(x) = a(y) \Rightarrow t(x) \neq t(y)$. That $a(D)$ is finite results from Lemma 1 and from the fact that we suppose that u is parallel to r if D has a ray. Suppose that $a(x) = a(y)$. From Lemma 1, there exists an integer k such that $y = x + kp_j u$. Moreover, since $x \neq y$, k is nonnull. Assume without loss of generality that $k > 0$. Then:

$$t(y) = \lfloor \lambda^T(y) - \alpha \rfloor = \lfloor \lambda^T x - \alpha + kp_j \lambda^T u \rfloor$$

but:

$$kp_j \lambda^T u = k\lambda^T u \left\lfloor \frac{1}{\lambda^T u} \right\rfloor \geq 1$$

which implies that:

$$t(y) \neq \lfloor \lambda^T x - \alpha \rfloor = t(x).$$ □

Appendix 9.3

Proof of Theorem 4

The proof of this theorem will use the following lemma:

Lemma 2 $a(x) = a(y)$ if and only if there exist integers k_1 and k_2 such that

$$x = y + k_1 p_j u + k_2 p_k r$$

Proof ⇒. Let $x = y + z$. We have:

$$\forall m \neq j, k; \ a_m(x) = a_m(y) + a_m(z)$$

which gives:

$$a_m(z) = 0 \tag{9.26}$$

On the other hand:

$$a_j(x) = a_j(y) + a_j(x) \bmod p_j$$

and thus, $a_j(z) \bmod p_j = 0$. Consequently, there exists an integer l_1 such that:

$$a_j(z) = l_1 p_j \tag{9.27}$$

Similarly, there exist l_2 such that:

$$a_k(z) = l_2 p_k \tag{9.28}$$

Solving the system given by (9.27) and (9.28) gives:

$$z_j = l_1 \frac{p_j}{\Delta} u_j + l_2 \frac{p_k}{\Delta} r_j$$

$$z_k = l_1 \frac{p_k}{\Delta} u_k + l_2 \frac{p_k}{\Delta} r_k \tag{9.29}$$

By substituting (9.29) in (9.26), we obtain:

$$\forall m \neq i, k: z_m = l_1 \frac{p_j}{\Delta u_m} + l_2 \frac{p_j}{\Delta r_m}$$

Since u and r are primitive, Δ must divide l_1 and l_2. Let $k_1 = \frac{l_1}{\Delta}$ and $k_2 = \frac{l_2}{\Delta}$. We then have $z = k_1 p_j u + k_2 p_k r$ which gives the result.
⇐ The sufficient condition can be easily proved by a simple calculation. □

Proof of Theorem 4

Let x and y be two points of D such that $a(x) = a(y)$. From Lemma 2, there exist k_1 and k_2 such that

$$y = x + k_1 p_j u + k_2 p_k r$$

Suppose now that $t(x) = t(y)$. We then have:

$$t(x) = [\lambda^T x - \alpha] = t(y) = [\lambda^T x - \alpha + k_1 p_j \lambda^T u + k_2 p_k \lambda^T r]$$

which implies that:

$$|k_1 p_j \lambda^T u + k_2 p_k \lambda^T r| < 1$$

But since $p_j \lambda^T u \geq 1$ and $p_k \lambda^T r \geq 1$ we must have:

$$|k_1 p_j \lambda^T u + k_2 p_k \lambda^T r| = 0$$

which implies that:

$$|k_1| p_j \lambda^T u = |k_2| p_k \lambda^T r$$

but,

$$|k_2| p_k \lambda^T r \geq p_k \lambda^T r \geq M(u) \lambda^T u + 1 > M(u) \lambda^T u$$

from which:

$$|k_1| > M(u)$$

and by definition of $M(u)$, this implies that x and y cannot belong simultaneously to D. □

Index